Structi

A. K. G WAY C

CONCRET

Its Composition and Use

A Clear, Detailed, Complete Statement of the Fundamental
Principles of the Basic Process of the Concrete Industry,
Including Essential Up-to-date, Proven Tables
and Data for the Users of Concrete.

*Embodying the complete text on this subject used
by the Institute of Applied Concrete Engineering*

By H. F. PORTER, C. E.
and Other Authorities, Associated with the Editors of
CONCRETE ENGINEERING

FIRST EDITION—SECOND THOUSAND

1909

A CONCRETE FACTORY BUILDING DURING CONSTRUCTION.

FOREWORD

This book is an endeavor to present in a direct way the underlying facts and the best present day practice in the basic process of the industry. The subject is developed through the history, character, analysis and manufacture of cement to a keen detailed analysis of the other materials entering into concrete. The analysis of sands and aggregates presents and discusses the problem intimately, vividly, directly and yet simply. Such treatment renders the problem easier.

The discussion of laboratory work in testing cement has been included because it presents in a strikingly clear and demonstrable manner, facts concerning Portland cement which perhaps could not be thoroughly grasped in any other way.

CONCRETE, ITS COMPOSITION AND USE, includes, as stated elsewhere, the complete text of the Institute of Applied Concrete Engineering, revised and augmented. The section on laboratory work (Chapter VI), especially shows the detail in which this instruction has been carried out.

The addition of the author's discussion of the colloidal theory of cementitious action, and Mr. Moyer's recently presented discussion of mineral oil in concrete, brings the matter down to the date of publication. While we feel that the book is going from our hands with many questions unsettled and many others suggested, yet it cannot be otherwise. Further development along these problems will throw more light on the subject and will be covered in later editions.

For the worker in concrete, and we all are to a greater or less extent, this volume includes a strong and carefully selected compilation of the best proven tables and data available. These are always of real value as guide and reference.

The Table of Contents has been omitted from the volume, but a carefully and practically prepared index should compensate for this. As a rule, subjects are not indexed under "cement" or "concrete." Look for the item directly. For instance, "specific gravity of cement" is indexed under "specific gravity."

Our earnest wish and endeavor is that this volume may help many to lay a foundation of intelligent comprehension of what concrete really is. and on which a higher and greater understanding can be built.

THE EDITORS.

Cleveland, January, 1910.

CHAPTER I.

CONCRETE, ITS COMPOSITION AND USE.

Lime, Puzzolans, and Natural Cement.

All Rock Formed by Cement.—The use of a cementing material for gluing hard, stony grains together to form larger masses of rocky substance is as old as the world's history. Nature, in every rocky formation, has shown the way; in sandstones, limestones, slates, shales and conglomerates, the tiny grains have become skillfully cemented one upon another until massive, rocky solids, even mountains, have resulted.

Man, keen to observe and analyze the whys and wherefores of all things, discovered the basis or principle of this accretion to be a chemical change, in which certain substances in solution, filtering through the loose grains, minute and great, of which the earth's crust is so largely formed, have become deposited in crystalline form, interlocking the particles as in a grip of steel.

Gypsum as an Example of a Crystalline Substance.—"Plaster of Paris" is gypsum minus the water of crystallization. It is a white powder composed of the elements calcium, sulphur, and oxygen chemically combined. Gypsum has in addition a certain amount of water in combination to which it owes its solid, semiopaque appearance. On heating sufficiently, this combined water is given off as vapor and the white powder "plaster of Paris" remains. Now on mixing this powder with water again, it hardens and reassumes its crystalline structure. Any gritty substance, admixed with this plaster paste, becomes locked in the structure. Here we have an example of a cementing process.

The action of lime is very similar—in fact this is the underlying principle of all cementitious action. Any variation or apparent difference is a matter of modification or refinement, in order to improve upon the extent and uniformity of this crystallization. Portland cement, as it is known to-day, is the most perfected form of cementing materials—the present final product of the evolution of the art.

Ancient Cements.—The familiarity of the ancients with cement is well attested by the condition of numbers of ruins extant in Greece, Italy, Egypt and Ireland, in which a cement was used, either as a mortar or as a concrete, that was good enough to withstand all the ravages of time. The dome of the Pantheon in Rome, built before the Christian era, is of concrete and is in good condition to-day. And about the Coliseum in Rome and the Appian Way, now scenic ruins, examples of concrete still intact abound.

, As to the precise nature of these ancient cements, we are ignorant. But some of it is supposed to have been made by grinding up lavas or volcanic slag, and for this reason slag cements to-day are often known as "Roman cements." While a satisfactory material, as evidenced by its endurance, tests of samples show that it was far inferior to the modern Portland cements.

Modern Cements.—Portland cement—so called because of a slight resemblance in color to a certain building stone quarried on the Isle of Portland in England, and called Portland stone, is indeed a product of evolution, and because of its relative superiority in quality and economy, is rapidly displacing all other cementitious materials. More than that, it is creating entirely new, ever-widening fields of usefulness, and is exerting a revolutionary effect on modern industry and progress. The present age may well be classified in history as the "Concrete Age."

Leading up to an exposition of Portland cements, their history, evolution, manufacture, qualities, and use, let us consider briefly the minor cementitious materials that still find usage.

The Principal Cementitious Substances.—The chief cementitious materials at the command of the builder to-day are, **Limes**, **Puzzolan** cements, **Natural** cements, and **Portland** cements. Some of these in turn fall into sub-classes which will be discussed under their respective main headings. These will be first defined and their qualities and use outlined, after which the details of manufacture and application will be discussed.

Limes.—Lime, more commonly called "quicklime," is the product of calcining certain limy rocks until their chemically combined water and carbonic acid gas are driven off.

The quality of the lime rock, or degree of purity, necessary to produce a lime that will slake readily will be given in a later paragraph, where a comparison of the composition of raw materials suitable for cements is given. Suffice it to say now, that the limestone must be very nearly pure calcium carbonate ($Ca\ CO_3$) in order to produce a freely slaking lime. This compound upon continued heating at a temperature of 700° to 900° Fahrenheit is broken up into Calcium oxide (CaO), which is the commercial product known as quicklime, and carbonate-acid gas (CO_2) which goes off into the air.

Upon adding water to the quicklime, a chemical change ensues, during which the lime slakes or dissolves, emitting considerable heat. The resulting product is a soft, putty-like substance which soon hardens or "sets," and if mixed with sand forms a species of artificial stone, which may be utilized to bond two stones or bricks together, making them as one. This is lime-mortar.

The compound formed on "slaking" is called lime hydrate, or calcium hydroxide ($Ca[OH]_2$), and it is the assumption of crystalline structure upon hardening that gives it cementitious value.

Lime, while widely occurrent, simple and inexpensive to prepare, requiring only a rock crusher and a plain kiln at the site of a limestone quarry, and although for many years the principal cementitious material, is being steadily supplanted by cements, to which in both strength and endurance, it is vastly inferior. Its use to-day is confined to the less progressive people and sections and to the poorest grades of brick and stone masonry; but even then, it is usually mixed with cement, which greatly increases its strength and permanence. In turn, lime imparts certain qualities to the cement, making a smoother working and more impermeable mortar; the maximum results, so far as workability, strength, and economy are concerned, are secured by combining cement and lime hydrate (in form of dry powder) in proportions approximately two to one respectively, and even equal parts have been found to give more satisfactory results than all of either one alone. Sabin (in "Cement and Concrete," 1905) reports the results of tests which showed an increase of 150 per cent. by substitution of one-half as much lime hydrate and 100 per cent. by substitution of an equal amount, besides greatly increasing the workability of the mortar. The effect of the lime hydrate, when used in connection with cement, seems to be to bring about a more complete combination of the silica present in the cement, otherwise uncombined and acting only as somuch sand. A fuller discussion of this phenomenon will be given in due course.

Hydraulic Limes.—Certain limes because of the presence of impurities in the shape of clay, after calcination, slake very feebly, if at all, when water is added, but soon after being mixed with water begin to harden. The resulting product is both harder and stronger than the ordinary lime mortar. To such is given the name hydraulic limes. This property of hydraulicity or hardening without slaking when mixed with water, is very important, and is the essential property of all cements.

Hydraulic limes resemble, on the other hand, the natural cements, and are often therewith classed. The difference is in the amount of free lime present; if there is enough to cause slaking, the product is a hydraulic lime; if not it is natural cement.

In Europe, about the middle of the 18th century, hydraulic limes began to be extensively used, supplanting the common limes very largely; but with the advent of natural and Portland cements their use began to decline until at the present time they are a waning competitor of natural cements only. In the United States their use has practically been confined to special purposes.

The appearance of clay with the lime, as an impurity in the limestone, is an important event, for it is the basis of all hydraulic properties, and the relative amounts of these two prime ingredients is the measure of the hydraulicity. This fact is made the basis of all analyses of raw materials to determine their character and fitness for cement making. It will be further discussed when the question of chemical composition of cements is taken up.

Cements Proper, Puzzolans.—Under this heading are commonly classed Roman or Puzzolan cement and Slag cement. The first is made by pulverizing a mixture of lava or volcanic tufa (a sort of natural slag) and lime; the second by pulverizing a mixture of slag and lime. The two are thus closely akin in composition and properties. The kind of puzzolan generally preferred for this purpose is known to the trade as **trass.**

Puzzolan cement is the kind supposed to have been used by the ancient Romans, and still finds extensive usage in the volcanic neighborhoods of Europe. It is a good cement, but inconsistently so, and neither so strong nor reliable as the natural cements. Because first used supposedly by the Romans, it is often called "Roman" cement.

Slag cement is made, as previously stated, of slag, or the refuse from the blast furnace, and is thus a by-product of the iron industry. To give the proper composition, a quantity of quicklime is added to the slag and thoroughly admixed by intimate grinding.

Ordinarily the slag issues in molten streams, which on cooling assumes a hard, glassy condition. When, however, it is wished to utilize the slag for making cement, the molten streams are run into big tanks, in the bottom and sides of which are innumerable little water or air spouts. The sudden cooling of the slag on striking these spouts causes it to be disintegrated into a granulated substance somewhat resembling coarse brown sugar. It is this substance intimately ground with a certain proportion of lime that is marketed as "**Slag Cement.**"

Slag cements are fairly strong, at times measuring up to Portland cement specifications, and if the percentage of sulphur present is low (it should not exceed 1½ per cent.) it is a safe cement to use for heavy mass concrete and other purposes where no great strength or wearing qualities are requisite. But, the fact that the percentage of sulphur varies and may be when least expected so high as to ruin the quality, and that there is no known method of economically insuring the removal of the sulphur, makes this cement unreliable. Moreover, the fact that the ingredients are only mechanically intermixed, not chemically combined, introduces a further factor of uncertainty and a dangerous one, for the reason that whether well or indifferently intermingled, the chemical analysis would give the same indications. Slag cement should, therefore, be used with caution and never for important work.

The distinguishing characteristics of Slag Cements are: Color, light bluish, shading to lilac; weight, (specific gravity) about 2.7 to 2.9 as against 3.2 for Portland, a barrel equalling about 330 lbs.; setting, slower than Portland; strength, lower than Portland; endurance, deficient in resistance to abrasion.

Slag Portland Cements.—There is a slag cement on the market called "Slag Portland cement." It is made by calcining the slag and lime in a kiln and then pulverizing the clinker into which the elements have thus been chemically combined. The process of manufacture resembles that of Portland cement, hence the generic. This cement is generally excellent, comparing favorably with the best Portlands, and should not be confounded with the ordinary slag cements. It will be dwelt upon more at length under Portland cements.

Natural Cements.—Cements made from the natural rock, without admixture of any other substance, are termed natural cements. The term does not signify a cement with either definite composition or characteristics, inasmuch as the composition of the cement rock not only varies according to localities but in the same deposit. The same brand of cement may differ considerably from itself as the quarry working advances. This fact makes the use of natural cements a difficult problem with which to deal in practice, and is one of the chief reasons why Portland cement is rapidly supplanting it. Nevertheless, natural cements have been of great service and will continue to be used for engineering work of lesser importance. Some, indeed, are favored with a composition naturally so like Portland that they are often substituted therefor.

Natural cements are a step nearer to Portland cements. In fact, some are almost identical in composition to Portland. Some, on the other hand, are little better than hydraulic limes. The difference lies chiefly in the proportion of free lime. If the excess is sufficient to give slight slaking properties then the product is a hydraulic lime; if after calcination and grinding the product shows no symptoms of slaking on addition of water, then it has passed the border line.

The distinguishing characteristics of natural cements are: Color, yellow to brown; weight, 2.7 to 3.1, a barrel weighing about 350 lbs.; setting qualities, sets more rapidly than Portland, but does not attain so high ultimate strength; strength, never so high as Portland cement; endurance, resistance to abrasion low, hence unsuitable for wearing surfaces. The chemical properties will be discussed elsewhere.

Historical.—The history of natural cements reaches back into the hazes of antiquity, antedating the Christian era. Their use continued in a small way down through the ages, but it was not until the 19th century that the world awakened to their value as a construction material.

In Europe, natural cements began to be manufactured on a considerable scale about 1800 and rapidly found favor, supplanting steadily the limes. During the past quarter century natural cements, in turn, have been giving way to Portlands.

In America, the history of natural cements begins with the year 1818 and owes its origin to the construction of the Erie Canal. It happened that a certain impure limestone intended for making lime for use in the masonry of the middle section of the canal, after burning refused to slake. Whereupon, one of the engineers, Mr. Canvass White, who had fortunately but recently visited Europe for the purpose of informing himself as to cements, suspected the true nature of the rock; and sure enough, upon grinding and mixing with water, the product developed a good set. The material was adopted for the work at hand and thus with the construction of the locks and walls of the middle section of the Erie Canal began the natural cement industry in the United States.

The description by the engineer in charge of this section of the Canal, Mr. Benjamin Wright, who with Mr. White made an exhaustive study of the material, "That it hardens best under water, and its properties seem to be partially lost if permitted to dry suddenly or if not used soon after mixing," express the characteristics of natural cement yet to-day as well as any definition.

The chemical composition of this first American natural cement showed about 83 per cent. impure limestone of which about 33 per cent. was magnesium carbonate and 16 per cent. clay, which would to-day classify it rather as a hydraulic lime.

The central region of New York State, comprising the counties west of Cayuga to Buffalo, became the center of the industry. Despite the wide distribution of suitable raw materials this district sets the pace, and with its neighbor, the Rosendale district in Rondout Valley at the foot of the Catskills, controls the bulk of the output. In quality as well as in quantity of product the New York districts maintained the lead for many years, but in the former it has more recently been shared with the Louisville district, Kentucky, producing Louisville cement.

Cement was discovered in the Rosendale district in 1825, also in excavating for a canal. The rock hereabouts was so well adapted to the purpose that the industry developed very rapidly until the Rosendale district led the country in both quality and quantity of the output. Indeed, the name Rosendale became synonymous with natural cement. Cement rock exists in this district in enormous quantities.

The production of natural cements, as shown in the reports of the United States Geological Survey on Mineral Resources, increased from a total of 300,000 barrels in the period 1818 to 1830, to a total of 78,000,000 barrels in the decade 1890 to 1900, an increase of many thousand per cent. The largest yearly output was in 1900, and since then the production, owing to the rapid increase in the use of Portland cement, has declined.

At the present time, the production of Portland cement has become so widespread, the standard of excellence brought so high and the cost so low, that the market for natural cements has been seriously undermined. Engineers now rarely, if ever, employ natural cements in works of importance and never in reinforced concrete construction. For heavy mass work and foundations, where no great strength is required, natural cements still meet with wide usage; but even then, only in favored districts, like the Rosendale for example, where the quality of the product is high and the economic conditions for manufacture and marketing are favorable, are the natural cements able to hold their own.

In consequence, the development of the natural cement industry in the United States is at a standstill; and in spite of the wide occurrence of suitable raw material, cheap fuel and favorable transportation facilities, wherever new plants are being installed the equipment is for the manufacture of Portland cement.

It is therefore evident that unless the quality of natural cements is improved, which it may be by introducing better grinding machinery and adopting means to refine and correct the raw material, their usage will decline until they are classed with hydraulic limes as a material of the past. The same mills will still continue to make cement, but the refinements competition will have compelled them to adopt will have evolved them into the class of Portland cement producers.

Manufacture.—The basis of natural cement production is an impure limestone, what may be called a clayey limestone or a calcareous shale, depending on the preponderance of the one or other of the essential ingredients, clay or lime. Alumina and iron are also generally present in small quantities, and magnesia as well, but in widely varying amounts. The essential ingredient of clay is silica.

The relative proportion of these compounds determines the degree of burning necessary and the properties of the product. This fact differentiates the process of manufacture in the degree of burning only.

The rock is either quarried or mined, depending on the nature and location of the formation. It is broken up into convenient sizes for handling by sledging and is then transported to the rock crusher at the site of the mill and crushed.

The crushed rock is fed into huge vertical kilns or furnaces, alternate layers of fuel and rock being introduced. Chemical changes begin to take place at 100°C. (212°F.), when the water is expelled. At 400° to 750°C. the magnesia breaks up, carbonic acid gas going off and magnesium oxide remaining. At 750 degrees to 800 degrees C., the calcium carbonate breaks up, and above 800 the clay breaks up, and aluminum, iron, calcium and magnesium begin to combine. Any alkalies present act as a flux. The silica is not attacked until 1100-1300°C. is reached, when lime and magnesium silicates are formed. Before this, only aluminates and ferrites have been formed. The larger the production of lime the higher the temperature necessary. If magnesia is high, replacing some of the lime, then evidently a lower temperature will do the work. Much depends upon the proper burning. In the chapter on chemical composition of cement material, the influence of the composition on burning will be further considered.

The burned material or clinker emerges from the bottom of the kiln and is generally taken as it is and ground, the grinding machinery being similar to that used in flour mills. If the clinker is high in free lime, some may be removed by slaking the ground cement, or better by steaming the clinker, when the disintegrating action of the slaking softens the material and reduces the cost of pulverizing. Excess of lime may also be removed by burning at a higher temperature, but manufacturers generally adopt the slaking method because cheaper and easier.

Many plants to-day are adopting rotary kilns and modern grinding machinery, such as are used in making Portland cement. The burning is thereby greatly improved, as well as facilitated and cheapened, and together with the greater fineness attained by using ball mills and tube mills, the quality of the cement is considerably improved.

The details of burning with rotary kilns and grinding with modern machinery being similar to the Portland cement processes, the description is reserved until Portland cement manufacture is considered.

Cost.—The following summary of the cost of producing Natural cement is given by Mr. Edwin C. Eckel, C. E., of the United States Geological Survey:

Rock at Mill.....................$0.05 to $0.10
Labor at Mill....................$0.06 to $0.17
Coal for Kiln....................$0.02 to $0.12
Coal for Power...................$0.02 to $0.05
Maintenance, Interest, Deprecia-
    tion, etc. ....$0.03 to $0.06
Total cost per bbl. of cement......$0.18 to $0.50

In which packing into barrels is not included.

Specifications.—The specifications for natural cements are not as comprehensive and definite as those for some other cements. The generally accepted standard, the report of the American Society for Testing Materials, for June, 1904, covers only low limed cements, and would exclude many, if not most American natural cements. A summary of these specifications is herewith given:

Definition.—The term natural cement shall be applied to the finely pulverized product resulting from the calcination of an argillaceous limestone at a temperature only sufficient to expel the carbonic acid gas.

Specific Gravity.—The specific gravity of the cement, thoroughly dried at 100°C. shall not be less than 2.8.

Fineness.—It shall leave by weight a residue of not more than 10 per cent. on the No. 100 and 30 per cent on the No. 200 sieves.

Time of Setting.—It shall develop initial set in not less than ten minutes and hard set in not less than thirty minutes nor more than three hours.

Tensile Strength.—The minimum requirements for tensile strength for briquettes one inch square in cross-section shall be within the following limits, and shall show no retrogression in strength within the periods specified.

NEAT CEMENT.

Age	Strength.
24 hours in moist air	50-100lb.
7 days (1 day in moist air, 6 days in water)	100-200lb.
28 days (1 day in moist air, 27 days in water)	200-300lb.

1 PART CEMENT, 3 PARTS STANDARD SAND.

7 days (1 day in moist air, 6 days in water)	25 to 75 lb.
28 days (1 day in moist air, 27 days in water)	75 to 150 lb.

Constancy of Volume.—Pats of neat cement, about 3 inches in diameter, one-half inch thick, tapering to a thin edge, shall be kept in moist air for a period of 24 hours. Another pat shall be kept in air at normal temperature, and a third kept in water maintained as near 70°F. as practicable.

These pats to be observed at intervals for at least 24 hours, and to be satisfactory, should remain firm and hard and show no signs of distortion, checking, cracking or disintegrating.

It will be noted that no tests for **Chemical composition** are given for the very obvious reason that the composition of the raw material is so varying that it would be impracticable to fix a definite standard.

CHAPTER II.

Portland Cement.

Definition.—There are several definitions of Portland cement, adopted by different engineering societies from time to time, but all converge to the one essential fact, that Portland cement is an artificial mixture, in certain well defined proportions, established by practice, of a number of common substances found in nature; and that the process of manufacture consists in intimately mingling the ingredients in the raw state by grinding, then chemically combining the various compounds into certain other compounds by burning at a high temperature, and lastly reducing the almost glassy clinker resulting from the burning to extreme fineness by again grinding. This resulting impalpable powder is what is known to the trade as Portland cement. Its composition and proportions will be discussed in detail in the section on "Composition."

Differentiation.—Vital points to the user of Portland cement are, wherein does it differ from other cements and how may it be known?

Comparison need only be made with Natural cements and slag cements, as the absence of slaking properties effectually distinguish both these cements and Portland from limes.

Portland cement, being a product fairly well standardized, in composition and method of manufacture, has certain fairly definite physical and chemical properties which enable its easy identification. In color, weight and fineness, period of set, soundness and tensile strength it presents more or less marked points of difference from other cements, so also in chemical properties.

Natural cement, being so variable in composition, presents correspondingly variable features. However, it may readily be distinguished by its yellow to brown color, lighter weight, greater coarseness, lower tensile strength and inability to stand the tests for soundness. Few samples of either natural or slag cement, mixed with water and setting for twenty-four hours, will stand immersion in hot water or steam for several hours without swelling, cracking and crumbling. This Portland cement is required to do.

Slag cement may be distinguished from Portland by its lighter color and weight, slower set and greenish appearance of the fracture of a pat. In tensile strength it is inferior, usually markedly so, and in chemical composition it is also readily distinguishable.

Identification.—The chief distinguishing characteristics of Portland cement are:

Color—bluish or steel gray.
Weight—specific gravity 3.10 to 3.20.

It is packed either in barrels or in bags, cloth or paper. A barrel contains the equivalent of four bags. A bag holds one cubic foot, measured loosely and averages in weight about 95 pounds. A barrel, hence, holds four cubic feet and weighs 380 pounds. It is uniform in appearance and has a feel like flour. If exposed to dampness it soon cakes up.

Setting Qualities.—Portland cement sets less rapidly than natural cements and faster than slag, but not always so, being influenced largely by the amount of gypsum added to the finished product by the manufacturers. If little or no gypsum is present, the cement may set up or harden in a few minutes. Normally for practical purposes, an endeavor is made to secure an initial set, or a degree of hardness enough to bear the end of a 1/12" wire weighted with one-fourth pound, in from 30 minutes to two hours, and a final or hard set, or ability to resist the impress of a 1/24" wire weighted with one pound, in from one to eight hours.

The setting test is at best only approximate, the amount of water and temperature influencing it considerably. For instance, a cement that with 25 per cent water and a temperature of 70 degrees .F. would take hard set in one hour, with 30 per cent water and the same temperature might require 1½ or two hours, and at 30 degrees F. might not harden for several hours.

Setting may be delayed by increasing wetness of mix, or by chilling, or agitating. In cold weather (32° F. and below) setting is very slow—a vital point to know; and thus for cold weather work a somewhat quicker setting cement should always be used, and the materials warmed. It is a vital principle that the setting once begun should be disturbed as little as possible—whether by tamping, treading upon, jarring, or by undue forcing of set by focusing of heat on any part, such as occurs in summer under the burning rays of the sun. If in frosty weather it is essential to protect the freshly laid surface with tarpaulins, straw, manure, etc., it is equally, if not more essential to protect it from the boiling rays of the sun by awnings or by covering with set sawdust or sand. Uniformity of conditions while setting is in progress is one of the most vital factors in securing a concrete of proper strength and enduring qualities.

Strength.—The tensile strength of neat or clear cement briquettes is used as a measure of the strength. Cement is seldom if ever relied on in tension, its virtue lying in its great compressive strength; but as the compressive strength is usually a fairly definite multiple of the tensile strength (from 6 to 10 times as much), the latter, because of its greater convenience in testing, may be used as the measure of the strength. A description of the test pieces or briquettes and methods of testing their strength is given elsewhere.

The tensile strength of Portland cement is vastly superior to that of all other cements. The accompanying diagram, adapted from Eckel's "Cements, Limes & Plasters," illustrates the superiority better than words can describe.

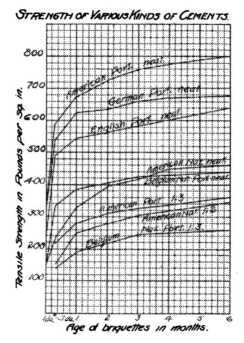

STRENGTH OF VARIOUS KINDS OF CEMENTS.

This diagram illustrates the relative value of the principal cements. These results represent the average of a large number of tests on leading brands of cement, made by the City of Philadelphia in 1897. They show the relative tensile strengths only, but as the compression strength is generally a certain proportion of the tensile strength, they indicate the true relative values very fairly. An interesting point to note is that one part of Portland cement with three parts of sand is better than one part of natural cement with only two parts of sand. Portland cement is thus more than one-third more effective than natural, and it would be more economical to pay $2.00 a barrel for Portland than $1.33 a barrel for natural.

The superior tensile strength of Portland cement is due mainly to three causes:

(1) To the perfect chemical combination of the ingredients.

(2) To the definite composition of the clinker.

(3) To the superior fineness of the cement (the finer the cement the greater its value).

The strength increases progressively for years, but at the end of six months it has practically its full value. For example, a cement may show a tensile strength of:

In 24 hours, moist air, 150 to 300 pounds per sq. inch, and even as high as 500 pounds.

In 7 days, 1 in moist air, 6 in water, 450 to 700 pounds.
In 28 days, 1 in moist air, 27 in water, 550 to 900 pounds.
In 6 months   700 to 1000 pounds.

The strength at 6 months is fully 95 per cent of its final strength. The strength at 30 days, under normal conditions, is fully 80 per cent of the final strength, so that a floor may safely be put in use when a month old. In winter, however, unless the setting is nursed by application of artificial heat, the strength at 60 days may still be insufficient.

In any records of tests, it is well to bear in mind that the results are always influenced by the personal elements. Differences in the amount of water, the temperature, thoroughness of mixing and thoroughness of kneading the soft cement into the molds, all vary the results. One manipulator may secure a strength of 500 pounds; another with the same cement may secure 600 or more. Statistics furnished by the mills should for this reason be taken "with a grain of salt," as they almost invariably will show abnormally high results.

A more valuable index of the strength of the cement is secured by the mortar test. In this, one part of the cement is mixed with two or three parts of sand. It is more valuable because it gives an idea of the sand-carrying capacity of the cement, the practical factor it is most desirable to know. Results, of course, vary with the nature of the sand; but to the end of uniformity of results, testers have adopted a standard sand which is a blend in definite proportions of grains of different sizes. Generally speaking, the cement with the largest sand-carrying capacity will yield the highest test in the 1 : 3 mixture, and is the best, although in the neat test it may not show up so well. A cement containing a proportion of coarse grains, due to insufficient grinding, which because of their size are inert and valueless as a cementing agency, may develop higher neat tests than cement of superior fineness. This is supposedly due to the increased adhesive surface afforded by the rounded modules of unground cement. These very particles, on the other hand, being the equivalent of so much sand, go to diminish the sand-carrying capacity of the cement, and thus to lessen the value of the cement for mortar or concrete.

Briquettes of mortar, proportioned one part cement to three parts of standard sand, should develop a tensile strength of:

In 7 days (24 hours moist air, 6 days water) 150 to 300 lbs. per square inch.

In 28 days (24 hours moist air, 27 days water) 200 to 500 lbs. per square inch.

A cement that shows up well neat but is weak in the 1 : 3 results should be questioned more closely than one of the opposite results.

Endurance.—Portland cement is especially well adapted for wearing surfaces or surfaces exposed to severe atmospheric conditions or chemical agencies. It is so fine naturally and so sound and hard when set up, that a well-trowelled surface of it as mortar will withstand ordinary wear and tear and the violence of wet and frost for centuries. All other cements differ in this respect—the presence of impurities, soft or underburned particles, imperfectly combined materials and injurious ingredients like sulphur and free lime constituting inherent destructive tendencies that unfit them for exposed surfaces.

The action of sea water on cement is not well established. There are examples of its resistance and also of its non-resistance to this agent. A French engineer, noting that cements high in alumina are generally most affected and those low in this element scarcely at all, has advanced the explanation that the disintegrating effect of sea water on cement surfaces is due to the expensive action of a certain gelatinous compound of alumina formed in the pores of the concrete by the interaction of the free lime, set free in the setting of the cement and magnesium sulphate, present in the sea water, which in turn reacts upon the alumina. If this is true, then the use of a cement as low as possible in alumina would go far toward solving the difficulty.

It has been shown that the denser the concrete, and hence the more impermeable the surface, the less the effect of the sea water.

By using, then, a cement of low alumina content and bending every attention to securing as dense and impermeable a facing as possible, the deleterious action may be greatly lessened if not entirely eliminated. By water-proofing the face exposed, the difficulty may also be met.

Strong acids, such as hydrochloric (muriatic) or sulphuric acid, will attack cement, but if the aggregate (sand and stone) be of silicious* and igneous† character, this action will be confined to a mere etching of the surface. In fact this means is often made use of to secure a surface of pleasing texture.

Oils in general have little or no injurious effect; nor have gaseous fumes, unless the aggregate is such as will be affected. Portland cement itself seems to be virtually immune to both oils and gaseous fumes.

Soundness.—The subject of soundness of Portland cement is one of the most important considerations, indicating as it does chemical purity and thoroughness of burning. Pats of neat cement, kept in moist air 24 hours, and then exposed to steam for 3 to 4 hours, or put in boiling water for the same length of time, should show no signs of swelling, curling up, cracking or crumbling. The presence of free lime or magnesia—in injurious amounts —is thereby revealed.

Chemical Properties.—The chemical properties of Portland cement concern the manufacturer and expert more than the user. Their detailed exposition is hence reserved for the chapter on Composition.

There are a few injurious ingredients and practice has established that:

(1) More than 1.75 per cent sulphuric acid (SO_3),

(2) More than 4.00 per cent magnesium or free lime, shall cause cement to be rejected for important use.

---

*A silicious sand is a quartz sand, composed of pure, or nearly so, SiO_2 (silicon dioxide).

†An igneous rock is a trap or basalt, the origin being volcanic.

The presence of any free lime whatsoever is injurious and ordinarily sufficient cause for the condemnation of the cement.

Gypsum is generally present, being added to the clinker (before grinding usually) to regulate the time of setting. The percentage should not exceed from 2 to 3 per cent, as more than this amount is an element of weakness.

The presence of free lime is disclosed, as stated before, in testing for soundness. It may also be determined by treating a little of the dry cement with muriatic acid, signs of effervescence indicating lime.

The amount of sulphur, free lime, or magnesia present may only be determined by quantitative chemical analysis and is a problem for the analytical chemist.

Discovery.—In the introductory section and in the section on natural cements, mention is made of the origin and development of Portland cement. It is pointed out that the modern Portland is the result of evolution, and represents to-day the fittest cement product. At this point some account of the history of the industry seems in order.

Portland cement grew out of natural cement. About the year 1824, it occurred to Joseph Aspdin, of Leeds, England, that if certain natural rocks when calcined and pulverized gave a cement, then an artificial combination of the same ingredients from different sources should give the same result. The fact that the quality of the cement could be thereby vastly improved, because of the opportunity afforded to secure a definite blend, perfected in every detail of manufacture, probably did not dawn upon Aspdin, but grew out of later experience.

Aspdin named his cement Portland because of a fancied resemblance to a certain limestone quarried on the Isle of Portland and known as Portland stone. The name Portland hence does not signify that Portland cement is or was made at Portland, England, Portland, Me., Portland, Oregon, or any other town of Portland, as has often been supposed. But the name has clung to the product from this early association, and is now so well established as to constitute authority for itself.

The first Portland cement was very little better than a natural cement, but the superior possibilities of the new process speedily became apparent, and by 1850 it had become firmly established. Its use spread rapidly throughout England and the continent, replacing steadily all other cements until to-day only a few brands of natural cement, like the Belgium, especially favored in composition, are able to compete.

In the United States Portland cement was first introduced about 1870 by David Saylor, and about 1875 the first works were established by him in Coplay, Pa. Other plants soon followed in this neighborhood. But progress was slow at the start, the new product having to face the usual prejudice confronting any new article. Engineers were reluctant to use it and continued to specify German or English brands in preference to the home brand for many years. Indeed, in some quarters, this early prejudice is not entirely overcome yet.

American Portland cement, however, soon began not only to equal but to excel foreign cements; and when through American ingenuity and enterprise the manufacture was practically revolutionized by the introduction of improved machinery and methods, the industry in the United States began to leap forward. By far the most important modification of the process was the introduction between 1888 and 1890 of the so-called rotary kiln, a horizontal revolving cylinder, instead of the old type vertical, stationary kilns, for burning the clinker. This considerably increased the output of the mill per horsepower and greatly reduced the cost per barrel. Along with the adoption of the rotary kiln came the introduction of improved grinding machinery, which likewise bettered the quality, increased the output and reduced the cost of production. These put American Portland cement right to the front, and gave it a lasting advantage.

The production from 1890 to 1900 increased more than 25 fold; and from 1900 to 1904 it tripled again. To-day there are Portland cement plants in practically every state in the Union and several in Canada, mostly in the province of Ontario and Quebec, and the statistics of the industry rank it well up in the list among industries. For rapidity of development there is no parallel in the industrial history of the world. What the future will be can hardly be imagined.

Evolution.—In the preceding section, it is pointed out that the Portland cement made by Aspdin was little better than a medium-grade, modern natural cement. Probably his cement would to-day scarcely pass the requirements for a natural cement. But of course improvement came with time.

When Portland cement was introduced in the United States, the German brands were in favor, a preference they maintained very generally until within a few years. To-day, however, the standard of American Portland has so far been raised by American ingenuity and enterprise that in quality and price it leads the world. More progress was made in America in a single decade than in the preceding 50 years abroad.

The increasing standard of excellence of American Portland cement has been due almost entirely to the adoption of superior methods and machinery. The composition has been modified, but not very much.

The early plants adopted the methods of manufacture in vogue in the natural cement industry; namely, the old-fashioned vertical kilns for burning, and ordinary flour-grinding machinery for pulverizing the clinker. As a result, the product suffered because of incomplete burning and inferior fineness of the final product.

The introduction of the rotary kiln, perfected at about the same time by two men working independently, Professor Spencer B. Newberry, of Cornell University, and David Durfree, about 1888 to 1890, marked a new era in the Portland cement industry.

A rotary kiln is simply a long steel cylinder (lined with firebrick), say 5 feet in diameter and 50 feet long, set on an axis concentric with its heads and inclined slightly from the level. The fine, crushed materials intermingled with powdered coal are fed in at the higher end, and as the kiln revolves the burning materials are steadily advanced toward the lower or discharge end, whence they emerge as completely burned clinker ready for final pulverizing. A fuller description of the rotary kiln is given in the section on manufacture.

The rotary kiln transformed the burning process from an uncertain and unreliable to a positive and more economical basis, greatly increasing the speed of handling, improving the quality and uniformity of the product, and reducing the cost of burning enormously. It gave the industry an impulse forward whose momentum cannot as yet be fully calculated.

The early rotary kilns burned oil. About 1895 Hurry and Siemans, engineers of the Atlas Portland Cement Co., perfected the kiln for using powdered coal as fuel. This method was a great gain over the old and was almost immediately universally adopted.

Thomas A. Edison, the wizard of the electrical world, was the next to introduce a revolutionary change. Until his advent in the field, the largest rotary kilns attempted were 5 to 6 feet in diameter and 60 to 80 feet long. Edison perfected a kiln 7 to 8 feet in diameter and 150 feet long. The output of the old kilns seldom exceeded 200 barrels per day; Edison's kiln was able to turn out 700 to 1,000 barrels per day. The new kiln was for some time regarded lightly by the Portland cement manufacturers, but soon won its way to recognition by sheer merit, and in 1905 all Edison's rights were absorbed by the American Cement Co., and the new kiln adopted. In 1908 the North American Portland Cement Co. acquired these and other patents. Edison's patents have been instrumental in further increasing the output and reducing the cost of production.

Along with improved burning came improved methods of grinding which have greatly increased the fineness economically possible and correspondingly enhanced the strength. A decade ago, 80 to 85 per cent of the cement passing as No. 100 screen (100 openings per lineal inch or 10,000 per square inch) was acceptable. To-day specifications are calling for at least 92 per cent to pass the No. 100 sieve and 75 per cent the No. 200 sieve (40,000 openings per square inch), and the best grades of Portland cement are surpassing these requirements. Formerly a cement pulling 500 pounds in 28 days was considered good; now a cement that does not show 500 in seven days is considered poor, and occasionally this strength is realized in 24 hours.

What the future will bring forth it is, of course, impossible to say, but it is safe to assume that betterment of the product will continue, and the cost be still further reduced.

The far-reaching effect of the development of Portland cement on the welfare and progress of the race can scarcely be overestimated.

The output of Portland cement in 1900 was 8,482,000 barrels; in 1906 it reached 46,000,000 barrels, and in 1908 the total reached the 55 million mark, and would unquestionably have exceeded even this had it not been for the business reaction. The increase thus, in the first seven years of the twentieth century, was in excess of 500 per cent.

Economics.—The raw materials out of which Portland cement is made are most widely distributed. The two essential ingredients, lime and clay, may be found together or close at hand in nearly every part of the globe. This fact differentiates the Portland cement industry from all other great industries born of the earth and rock or of the mineral deposits therein, such as the iron and coal industries, workable deposits of which occur only in a certain few localities. As a result cement mills are springing up in many parts of the United States, responsive to the demand, present and prospective. The far-reaching economic portent of this fact may well be realized. It renders a trust or monopoly of the output, with all its possible evils of high prices, inferior quality, etc., exceedingly difficult; it means a minimum addition to the cost of production due to haulage; the condition of a local supply for a local market makes every locality independent, eliminating the evils of interstate or foreign trade, jealousies, rivalries, tariffs, and the cost in time and money of long hauls. But degeneration is forestalled by the healthy competition insured by the overlapping of local fields of production. All this means a marked gain to the wealth and welfare of each community at the expense of no other, and plants the industry itself on a rock basis of confidence.

Composition.—It should be noted that the raw materials, while most widely distributed, occur in many different physical forms, so that often the excessive cost of preparing a certain deposit of raw material, suitable in chemical composition, may make it unavailable; or the deposit may present such a variable composition that its use may occasion constant worry, or may contain injurious impurities that unfit it for use. There are, for instance, large deposits of very hard limestone, which chemically are ideally suited for the purpose, but at present are practically out of consideration on account of the expense of grinding. Improved grinding machinery will eventually make it economically possible to use these deposits. There are also many deposits of limestone (known as dolomite) containing magnesium which are not being used at present because magnesium is considered deleterious to the quality of the cement; but it is possible that a modification of the process of manufacture may produce a good cement of magnesium, for magnesium and calcium (the principal part of limestone) are very much alike in chemical properties. If so the available supply of suitable raw materials will be still further increased and the industry correspondingly broadened.

Limestone is not the only source of lime, nor clay beds of clay: marls, chalks, and calcarious by-products of certain chemical processes (such as caustic soda and ammonia) furnish lime; and shales and slates, which are merely clays solidified by natural action, are used instead of clay. Blast furnace slag also furnishes suitable raw material.

These materials are so common and so abundant that an almost infinite variety of combinations of raw materials would seem possible to produce Portland cement. Practical considerations, however, reduce at present the possible combinations to a very few. In the United States typical combinations are

*(1) Clayey hard limestone plus clay (or shale)
(2) Pure hard limestone plus clay (or shale)
(3) Soft (Chalky) limestone plus clay (or shale)
(4) Marl plus clay (or shale)
(5) Alkali Waste plus clay (or shale)
(6) Slag plus pure limestone.

If the proportion of clay in the limestone is sufficient, or in other words if the rock naturally contains the proper balance of materials to produce a Portland cement, nothing need be added. But usually rock of this character will vary in composition as the quarry working advances, so that admixture of either pure limestone or clay in varying amounts becomes necessary, unless it be desired to produce only a Natural cement. Such is the case with the Rosendale and Buffalo-Akron districts in New York State and the Louisville district in Kentucky. The Rosendale rock particularly is almost a perfect composition, and the cement produced ranks up as high as some Portlands. Better burning and finer grinding would to-day make these Rosendale cements equal to some of the best Portlands, but perhaps not consistently so, on account of the bound-to-be variability of the natural rock.

An example of the use of the first combination is the "Lehigh" district in Eastern Pennsylvania. Here an outcrop of clayey limestone, known in geology as the "Trenton limestone" and to the cement world as "Cement Rock," extending from Stewartsville, N. Y., to near Reading, Pa., is found. This is almost ideal for cement making; it is easily quarried, easy to grind and in composition is so well balanced that only a small amount of pure limestone has to be added. It was in this district, at Coplay, Pa., that the first Portland cement mill in America began operation about 1875. To-day in a district little over four miles wide and 125 miles long from N. E. to S. W. across the State of Pennsylvania, there are over 20 Portland cement mills supplying over half the total production in the U. S.

Underlying the "Cement Rock" is a formation known as Kittanning limestone. This is a limestone relatively free from clay and containing more or less magnesia. Where sufficiently low in magnesia, it is used for the pure limestone needed to correct the "cement rock"—otherwise it is valuable only as road metal for

*(Authority—Edwin C. Eckel, C. E., U. S. Geological Survey.)

blast furnace flux, although it may one day become available for cement (as pointed out above). This formation is light gray to light blue in color and inclined to be cherty and is thus readily distinguished from the overlying cement rock which is slaty in nature and blackish in appearance.

In the neighborhood of Nazareth and Bath, Pa., the cement rock runs so high in lime that instead of additional pure limestone being needed, it is necessary to add clay or shale. The Trenton formation, in a purer form, continues to outcrop down into Maryland, Virginia, and West Virginia, and upward into Orange Co., N. Y., so that it is to be expected that Portland cement plants will spring up in all these regions.

Examples of the use of marl and clay are found in Ohio-Michigan districts and in Ontario and Quebec, also in the southern states, and in Iowa, Kansas, Nebraska and California. Examples of the use of hard limestone and clay are in Indiana and Illinois, where the Bedford limestone becomes available. In the region west of the Mississippi most of the mills are of this character. The most conspicuous example of the use of slag to produce a genuine Portland cement is found in South Chicago where the U. S. Steel Corporation is operating immense mills for the special purpose of utilizing the slag from its blast furnaces located there. In the neighborhood of Pittsburg are found similar producers.

Marl users are at a growing disadvantage because of the added expense of drying their material before they can grind it. Marl contains a great deal of water, and the cost of evaporating this is high. Consequently we find to-day, that the use of marl is decreasing, while the use of hard limestones is increasing. Soft limestones have not been resorted to much as yet, but present possibilities augur a future for them.

For the benefit of those who are not familiar with the terms for raw materials cited above, or are not familiar with geology, the following glossary is appended.

Limestone.—A dense, hard rock, usually grayish or bluish in color with a conchoidal fracture (not in layers). When pure it contains nothing but carbonate of lime or calcium carbonate ($Ca\,CO_3$), but there may be intermingled with it varying amounts of clay, which alter its color and texture markedly—the more clay the darker in color and the more slaty in fracture. If more than 50 per cent is clayey the rock is better known as a limy or calcarious shale. If magnesium is present, then the rock is called a magnesium limestone or dolomite.

Soft or Chalky Limestones.—Limestones either so finely divided in composition—as pumice—that they are soft and friable or so saturated with water that they have never had opportunity to harden. The chalk cliffs of England are an example of the first. Deposits of chalky limestone occur throughout the southern states, from Alabama westward to the Mississippi river.

Marl is almost pure calcium carbonate. It is soft and cheesy in appearance because supersaturated with water. When devoid of vegetable impurities it is quite white in color. It is found in the basins of dried-up lakes or swamp regions and is supposed to be made up of accumulated precipitations of lime carbonate held in solution in the water, and the skeletons of minute organisms that once lived in the water. The vegetable matter imprisoned is harmless as it burns up when the material is put in the kiln. Of course, its presence occasions extra cost in handling and burning, which is the objectionable feature.

Alkali Waste: Calcarious matter resulting from the manufacture of certain chemicals, notably caustic soda and ammonia. The by-product of the former is not generally so reliable as the latter as it may contain sulphur in injurious amounts, whereas the calcium carbonate from the manufacture of ammonia is usually quite pure. There are very few cement mills relying on this source for their lime.

Slag: Slag is a by-product of the iron industry. It is the refuse that is drawn off the blast furnace, leaving the pure molten iron in the cauldron. When allowed to cool directly it becomes a hard glassy mass, very tough and durable. But when it is desired to dispose of it handily or to utilize it, the molten streams are run into huge cisterns from the bottom and sides of which play innumerable little streams of air and water. The molten slag impinging on these is converted into a granular substance somewhat resembling coarse brown sugar in appearance. It is in this form that it is used as a basis of Portland cement. It is a compound of impurities in the iron ore and the limestone used as a flux in the blast furnace and in composition is like a calcarious shale—i. e. contains a superabundance of clayey matter, so that to correct it for cement use, pure limestone must be added. The objectionable feature of the use of slags is the liability of the occurrence of injurious amounts of sulphur.

Clays: These are variously clay, shale or slate, depending on the degree of solidification. Clays result from the erosion of rocks and comprise a large part of the earth's crust. They are composed of alumina ($Al_2 O_3$), silica ($Si O_2$) and iron oxide ($Fe_2 O_3$). To be suitable for the manufacture of Portland cement, they should be free from coarse sand and gravel, chunks of hard flinty rock, iron sulphide (pyrites), magnesia and alkalies, as the presence of these in any quantity introduces undesirable complications in the manufacture. To be most suitable, they should carry not less than 55 per cent silica and better, 60 to 70 per cent. The alumina and iron oxide together should not amount to more than half the silica. The nearer the ratio, $Al_2 O_3 + Fe_2 O_3 = Si O_2 \div 3$, is approached, the better suited the clay.

Shales are not so desirable, as their proneness to occur with more or less limy matter complicates the regulation of the composition, making the product more costly and less reliable.

Slates are not being used to speak of at present, but offer promising possibilities. Especially, as an adjunct to a slate quarry, would slate utilization be valuable, for fully 75 per cent of the slate quarried goes to the scrap pile, all of which could be utilized as raw material for cement.

CHAPTER III.

Chemical Composition of Portland Cement.

Similarity of Cements.—In preceding chapters, the close relationship between limes and cements is pointed out. Pure limestone, when calcined, gives the product quicklime. Adulterated with a small amount of clay it gives a **hydraulic lime**, or a lime which hardens or sets up when mixed with water (the product still slaking). When the proportion of clay present is sufficient to antidote all slaking, there is a marked increase in the hydraulic properties of the product, or the degree of hardness and strength it is capable of assuming. The product is then a true cement, and in a Portland cement the proportions of clay and limestone are so adjusted artificially as to give a cement of maximum practical value.

An ideal mixture for Portland cement would be (according to Spencer B. Newbury, Sandusky Portland Cement Co.),

75% Calcium Carbonate (Ca CO_3)
20% Clay (SiO_2, Al_2O_3, Fe_2O_3)
5 % Magnesia, Sulphur and Alkalies.

A theoretically ideal Portland cement would be pure **Tricalcium Silicate, (3CaO) (SiO_2)** and would then contain:

73.6% Calcium Oxide (**CaO**)
26.4% Silica (SiO_2).

For explanation of these percentages see Chapter Four.

Adulterants, unavoidably present in the raw material and some of them very useful adjuncts in facilitating the manufacture, prevent this theoretical ideal being realized. The practical ideal first mentioned is often closely approximated.

In the following table are given for comparison typical analyses of the principal cementitious materials of the Lime family. Note the effect of the variation in percentages of the different constituents on the classification of the product. The natural cements are very similar to the Portlands, but their liability to variable composition, as pointed out in the chapter on natural cements debars them from the Portland class. In the last line of the table is given the **Cementation Index**, the meaning of which will be given in the following chapter. It should be noted that all chemical analyses are reported in percentages; i. e. 50 in the table means that half of the product is of that compound.

TABLE SHOWING VARIATION IN COMPOSITION OF CEMENTITIOUS MATERIALS.

CEMENT CONSTITUENTS	LIME		HYD.LIME		NATURAL CEMENT				SLAG CEM.	PORT CEM.
	1	2	3	4	5	6	7	8		
Silica (SiO_2)			21.2	22.02	22.17	26.0	21.10	28.13	28.8	21.82
Alumina (Al_2O_3)	trace	trace		1.29	4.60	6.0	6.44	7.37	12.0	7.53
Iron (Fe_2O_3)	trace	trace		1.00	1.23	4.0	2.66	1.73		3.50
Lime (CaO)	90-95	55-90	78.8	74.51	60.06	35.00	39.70	43.79	51.2	62.50
Magnesia (MgO)	0-5	6-42		1.18	0.73	20.0	22.90	10.43	2.64	2.24
Alkalies (K_2O, Na_2O)					3.0	2.50	2.22			1.24
Anhy.Sulphuric Acid (SO_3)						1.2	1.35			1.20
Carbon Dioxide (CO_2)	trace	trace			1.46	2.0	1.35	9.28	4.0	
Water (H_2O)	trace	trace			0.48	2.7	2.00			
Sulphur (S)									1.4	
Cementation Index			0.75	0.85	1.09	1.30	0.91	1.26	1.66	1.09

1. High Calcium Lime, containing less than 5 per cent magnesia.
2. Magnesium Lime, containing more than 5 per cent magnesia, generally about 30 per cent.
3. An ideal composition for a Hydraulic Lime.
4. Hydraulic Lime from Teil, France (average of 7 analyses).
5. Belgium Natural cement (representative analysis).
6. Rosendale Natural cement (representative analysis).
7. Akron-Buffalo Natural cement (representative analysis).
8. Louisville Natural cement (representative analysis).

The percentages for Portland cement are the average of 80 standard American Brands.

The trade name, chemical term and symbol of the various compounds met with in dealing with cement, are appended for convenient reference at the end of this chapter (page 33).

Also for the sake of reference, making the table mentioned more intelligible, and for use in calculating the chemical reactions that occur in the manufacture of cement, a table of the elementary substances figuring in cement compounds is herewith given. The name, symbol and atomic weight or ratio of heaviness based on hydrogen as unity, appear:

Chemical Name.	Symbol.	Atomic Weight.
Aluminum	Al	26.9
Calcium	Ca	39.7

Chemical Name.	Symbol.	Atomic Weight.
Carbon	C	11.9
Hydrogen	H	1.0
Magnesium	Mg	24.2
Manganese	Mn	54.6
Nitrogen	N	13.9
Oxygen	O	15.9
Phosphorous	P	30.8
Potassium	K	38.9
Silicon	Si	28.2
Sodium	Na	22.9
Sulphur	S	31.8
Titanium	Ti	47.7
Iron	Fe	55.5

To those unfamiliar with chemical terms, it may be explained that practically all substances as found in nature are compounds of various elements, and are combined in certain definite proportions. Elements are held to be substances no further separable.

Examples of elements are gold, lead, carbon. Common salt is not an element but a compound, composed of the elements sodium (Na) and chlorine (Cl). Each element has a characteristic weight, such that a cubic foot weights a certain definite amount. A cubic foot of pure iron, e.g., weighs in the neighborhood of 480 pounds.

For the purpose of simplifying calculation of chemical reactions, an arbitrary scale of weights has been chosen, called the table of atomic weights. The most common standard or unit is the element hydrogen (H), the lightest known element. A volume of hydrogen under the same conditions of temperature and pressure weighs only 1-16th as much as an equal volume of oxygen (O), and so on. Comparison is based on a standard temperature of 70°F. and ordinary atmospheric pressure.

In calculating the combined atomic or molecular weight of a substance, as e. g., SiO_2, the atomic weight of Si is added to twice the atomic weight of O (the subscript indicates the number of atoms or weights of the element appearing in the compound). Thus:

$$\text{At. Wt. Si} = 28.2$$
$$2 \times \text{At. Wt. O} = 31.8$$
$$\text{Molecular Wt. } SiO_2 = 60.0$$

Certain of the elements when combined with oxygen (O) are acidic in their reaction; others basic or alkaline. The reaction of an acidic substance and a basic yields a salt or neutral substance. Salts are among the most stable compounds known. An example of a salt is sodium chloride (NaCl), or ordinary table salt. Pure Portland Cement, when hydrated, is a salt and a very stable one.

Its ideal composition, as given on page 26, would be pure Tri-Calcium Silicate $(3CaO.SiO_2)$; in which the CaO is the alkaline constituent and the SiO_2 the acidic. The water also plays the part of an acid. The acids and bases equalizing one another produce a substance of a high degree of neutrality, hence the stability of Portland cement and its high enduring properties, which makes possible with this material practically everlasting structures.

Lesser Compounds Present.—Alumina is generally present, forming with part of the lime the dicalcium aluminate (2CaO) (SiO_2). Iron oxide is also generally present and is held to act like the alumina, so that the two may be coupled in calculating the proper proportions. There are also other impurities, such as magnesium, sulphur and alkalies unavoidably present, and adulterants as calcium sulphate (gypsum) intentionally added for the purpose of regulating the time of setting. More than 1.75% sulphur and more than 4% magnesium are considered injurious, and are sufficient to reject the raw material. Materials containing more than 5% alkalies should also be regarded with suspicion if not rejected.

Maximum Practical Value.—Generally speaking, the higher the amount of the tricalcium silicate produced in the operation of burning, the better the cement; and if a clinker of pure tricalcium silicate were to be produced the resulting cement would test very high. But to fuse pure CaO and SiO_2 into a clinker would require an extraordinary heat, such as could be produced only in an electric furnace or with the oxy-hydrogen blowpipe, and the clinker would be so hard as to require special grinding machinery. The more alumina and iron oxide present the lower the clinkering temperature but the less the value of the product. Hence it becomes necessary to strike a balance between the disadvantage of a weak cement and the advantage of a low clinkering temperature. This condition will be further discussed elsewhere.

Recapitulation.—The *normal constituents* of Portland cement are: *Lime, Silica, Alumina* and *Iron Oxide.* Common impurities are *Magnesia, Sulphur* and *alkalies.* Ideally pure Portland cement would contain only *Lime* and *Silica* and its formula would be (3 CaO) (SiO_2), the chemical combination of the stable basic oxide and the stable acid oxide being effected by a high degree of temperature. As the degree of temperature required for this fusion is too great to be practically available at the present time resort is had to certain adulterants which act as a flux and permit the fusion at a degree of temperature practically obtainable. Alumina and iron oxide conveniently present in most clays along with the silica, serve this purpose admirably. These two compounds are generally considered interchangeable in their effect on the cement, and one or the other may be entirely absent. The other substances liable to be present in small percentages are harmless to the product, serving merely as inert impurities. Most of the sulphur, unless present as a sulphide, passes off as the gas, SO_3, in the kiln, and the alkalies, too, are sublimated (disintegrated and set free as fine dust) during this stage of the manufacture.

It is important that the user of Portland cement should be aware of the chemical composition of the cement he is using and know the effects of each constituent on its quality and behavior. At the risk of repeating some points already covered then, the following itemization of the compounds commonly met with in analysis of cement and the influence of each, is given:

. **Lime (CaO)**—As previously noted, a theoretically perfect cement would contain 73.6 per cent. lime (calculated as oxide and not carbonate). To attempt this proportion is, however, impractical, and ordinary cements show nearer 60 per cent. lime. Of course, theoretically, the closer the lime content approaches the ideal percentage the better the product; but as it is impracticable to realize perfect mixing, grinding and calcining, the attempt to carry the lime too high would be disastrous, for it would be impossible to combine it all, and in the uncombined state it would be a source of weakness. Its presence in the finished product is manifested in swelling and disintegration, and, if not the ruin of the work, at least its unsightly discoloration. The aim of the manufacturer is, then, to keep the lime content down to the point where he may be sure all of it is combined. This may result in an excess of clayey matter, but further than reducing the tensile strength of the product slightly, because the clay replaces just so much sand, it is harmless. There is on the other hand the danger of carrying the lime content so low that the tricalcic silicate is only partly formed, the rest being the incomplete dicalcic silicate, $(2CaO)-(SiO_2)$, which crumbles to powder. The result is the clinker falls to pieces in the kiln and the product is worthless.

Manufacturers, in placing a new brand on the market, in order to secure a very high testing cement, will often carry the lime content higher than commercially practical, and will then for a time, or until the product has won its way to favor, spend extra time and expense in intimately mixing, grinding very fine and burning it very hard in order to ensure the combination of all the lime. Then, when the product is established, they drop the lime content to a more economical percentage.

The better the grinding machinery and the better the burning apparatus, the higher the lime may be carried, and the better the cement.. This explains why steady improvement in grinding machinery and the replacing of the old vertical dome kiln by the horizontal rotary kiln and its gradual perfection, has evidenced itself in a constantly rising standard of excellence of American Portland cements.

Silica (SiO₂).—The theoretically perfect cement would contain 26.4 per cent. silica. In practice some of the silica is usually replaced by alumina and iron oxide, and its percentage may vary from 19 to 24 per cent. Obviously the higher the silica, if it can be combined, the better the cement, but the more silica, the less alumina and iron oxide can be present and the greater the difficulty in clinkering. Free or uncombined silica is harmless to soundness of the cement, constituting only so much inert matter.

Alumina (Al₂O₃).—This compound, a normal constituent of most clays, enters into the composition of practically all cements. Its chief function is to serve as a flux to facilitate the combination of the lime and silica. Itself combines with part of the lime to form the dicalcic aluminate (2CaO.Al₂O₃). The effect of the alumina on the process is to lower the clinkering temperature; on the product, it is to quicken the set, thus requiring the addition of an increasing amount of gypsum to slow the set, and it also tends to reduce the ultimate strength attained. Too much alumina in the raw materials makes the clinker too fusible and prone to ball in the kiln; too little means greater difficulty in clinkering. The less alumina in the product, the better the cement—an excess means an unreliable and perhaps hazardously quick setting cement, and low ultimate strength, both undesirable. Moreover, as pointed out previously, a cement high in alumina is practically worthless for salt water usage and for this purpose a cement as low as possible in alumina should be adopted. According to Le Chatelier, who has formed his opinion from experiment and analysis, this deteriorating effect is due to the interaction of magnesium sulphate, present in the sea-water, and free lime, liberated in the hardening of the cement, forming calcium sulphate, which in turn reacts with the alumina present to form **gelatinous calcium sulpho-aluminate.** It is to the swelling action of this last compound that the disintegration is supposedly due. Now as some fresh waters may also contain magnesium sulphate, this danger is liable to be met with in the case of a cement high in alumina in river or lake work, or even in any case if the water used in mixing should contain magnesium sulphate or calcium sulphate. The same danger presents itself if it has been necessary to add considerable gypsum (calcium sulphate) to a cement high in alumina, to retard the set; for with the addition of water, the same expansive compound would be formed. Cements, therefore, of this character should be used with caution.

Iron Oxide (Fe_2O_3).—This is ordinarily present in very small quantities and is then treated as so much alumina, allowance of course being made for the difference in molecular weight. There are cements with no Fe_2O_3 and there are also good cements in which this compound is present to the exclusion of alumina. One important difference exists, in that the iron does not act like the alumina as above, to introduce a disintegrating element, nor does it act appreciably to hasten the set. Such a cement, therefore, would seem especially adapted for sea-water use.

This completes the normal constituents. It is to be observed that the virtue of a cement is due to the amount of lime and silica combined and that the alumina and iron oxide act merely as a flux and otherwise are more detrimental than useful. Experience has shown that there will be enough of these adulterants present to properly assist the fusion and yet not seriously affect the value of the product if there is approximately three times as much silica as alumina and iron together, i. e. if

$$\frac{SiO_2}{Al_2O_3 + Fe_2O_3} = 3.$$

Miscellaneous Adulterants.—The influence of the other substances liable to be present in the raw material, has already been dwelt upon, and the limiting percentages given. Also, in the chapter on "Raw Materials" it was pointed out that inasmuch as the element magnesium is akin to calcium, forming similar compounds and exhibiting similar properties. it is perfectly possible that a line of magnesium cements will one day be evolved; but these will probably differ enough in method of preparation and in final properties to constitute a class by themselves, and Portland cement proper remains always in the "lime family." The small percentage of magnesia present in Portland cements is regarded in its effect as so much lime.

Following is a table showing the trade name, chemical term and symbol of the various compounds met with in dealing with cement:

Trade Name.	Chemical Term.	Chemical Symbol.
Lime (Quicklime)	Calcium Oxide	CaO
Calcite	Calcium Carbonate	$CaCO_3$
Dolomite	Calcium-Magnesium Carbonate	$CaCO_3 + MgCO_3$
	Tricalcium Silicate	$3CaO + SiO_2$
Slaked Lime	Calcium Hydroxide	$Ca(OH)_2$
Dead-burned Plaster	Calcium Sulphate	$CaSO_4$
Gypsum	Hydrous Calcium Sulphate	$CaSO_4 + 2H_2O$
Magnesia	Magnesium Oxide	MgO
Water		H_2O
Magnesite	Magnesium Carbonate	$Mg\,CO_3$
Quartz	Silica	SiO_2
Iron Rust	Ferric Oxide	Fe_2O_3
Iron Rust	Ferrous Oxide	FeO

CHAPTER IV.

Balancing of the Compounds and Calculation of the Mix.

Stability of Portland Cement.—As pointed out previously, the extraordinary stability of Portland cement is due largely to the fact that the acidic and basic (or alkaline) constituents of the raw materials are combined in the kiln in such a manner that the ground product, on addition of water, enters at once upon a peculiar chemical change, whereby a highly neutral jelly-like substance is formed. This gel is the true Portland cement and in its subsequent action, whereby hardening ensues, acts like a glue, except that unlike ordinary glue, which is of an animal or vegetable órigin and soluble, it is mineral and insoluble; and being insoluble, goes ahead with its work even though immersed. Hence the term hydraulicity, or property of hardening under water, applied distinctively to cements. If the combination of materials is improper, the formation of gel may be interrupted and instead, compounds of an inferior nature encouraged, decreasing the cementitious value.

By inferior nature is meant, of a character less efficient as a binder. The formation of solids from a solution, (matter ordinarily solid in chemical combination with water) takes place in three ways; precipitation in the form of large, well-formed, generally hexagonal crystals, from a weak solution or a cold one, taking place very slowly; precipitation in the form of pine-needle like crystals, from a medium solution or a weak solution warm, and precipitation in the colloidal (unbroken structure) condition in the form of a gel, from a strong solution or a moderate solution very warm. *It is to the collodial formation that Portland cement owes its virtue chiefly; thus any condition that disfavors the formation of colloids is detrimental to the strength of the cement.* The principal colloidal substance formed by the interaction of cement and water is a semi-solid solution of silicates (also aluminates and ferrates) in lime, or **calcium hydro-silicate,** which gradually, where intermingled with a mass of inerts, assumes an entirely solid condition, interlocking all the particles and thus forming a solid block, which is the concrete. Now, if the proportions of the raw materials are improper, or the calcium incomplete, or any other imperfection ensues during the preparation, the formation of compounds of a lower order, assuming a more or less unstable crystalline state instead of a stable colloidal, is induced; whence the inferiority. The exact proportion of compounds and thorough fusion in the kiln are necessary to induce an incipient chemical change which can be reassumed in the final mix with the greater surety of its proper and speedy completion. The presence of compounds alien to the correct formation, or of an excess or lack of one or other of the essential ingredients, manifestly militates against success, either by way of unsoundness or reduced strength.

In the case of Portland cements, conditions are made more or less a scientific certainty by exact balancing of ingredients; in the case of Natural cements, proper conditions are a happy occurrence, owing to the variable composition of the Natural cement rock; hence the reliability of the former and the uncertainty of the latter for important usages. In the case of hydraulic limes. the balance is so imperfect that free active agencies are always present, lowering the value of the product as a cement. The chief free active agent is lime, **CaO**, which must be slaked out before the product can be used; the resulting compound is a mixture of hydrated lime and cement, whence its hydraulic properties. In the case of ordinary lime, the chemical combination at the basis of the formation of cement is altogether absent, and the value of the product is due not to any hydraulic properties, but to the carbonate condition (limestone) by absorption of carbon dioxide from the air. Hardened lime mortar is thus nothing more nor less than limestone again—its original condition.

There is another way of looking at the change that takes place in the kiln, namely, that it is not a chemical change at all, at least it is only partially so, but a dissolving process by which silica, iron and aluminum enter into solid solution with lime and magnesia. The real chemical change, by which Portland cement is converted into a very stable mineral glue-like substance, the true cement, is postponed until the water is added. The water, then, in combination with the silica (and analogous compounds) acts to counterbalance the lime, inducing the requisite neutrality; there would be required, assuming the theoretical composition already noted, about **6 parts water (H_2O)** to bring about this change. This relation expressed chemically would read:

($3CaO.SiO_2$) plus $6H_2O$; or with the ions and anions (acids and bases) paired, **3 CaO** plus ($SiO_2.6H_2O$).

The molecular weights of either member of this relation will now be found equal (about) which is the condition requisite for neutrality or stability.

Assuming the formation of tricalcic silicate to be ideal, a fundamental relation can evidently be established on the basis of which cements can be compared and rated. The ratio or measure is most universally assumed as unity. Expressing the ideal compound in this form, it is found that the silica must be increased 2.8 times, that is, 2.8 times per cent. SiO_2 equals per cent. CaO. This may be called the measure of neutrality or stability.

So the relative departure from the ideal composition will be shown by the variation in this measure or ratio. This fact is made use of practically in almost every analysis of raw materials, and at all stages of manufacture, and is in fact the guide used in proportioning the mix and grading the product. For convenience the ratio thus established is called the "Cementation Index." It will further explained elsewhere.

CONSTITUENT COMPOUNDS.

In a normal cement, the compounds formed in burning are probably as follows:

Tricalcic Silicate ($3CaO. SiO_2$) (pure Portland cement).
Dicalcic Aluminate ($2CaO. Al_2O_3$).
Dicalcic Ferrate ($2CaO. Fe_2O_3$).
Trimagnesic Silicate ($3MgO. SiO_2$).
Dimagnesic Aluminate ($2MgO. Al_2O_3$).
Dimagnesic Ferrate ($2MgO. Fe_2O_3$).

Hence it will be seen that more than the acidic compound, SiO_2, and the basic compound, CaO, must be taken into account, in calculating the Cementation Index. The acidic compounds present are the silica (SiO_2), iron oxide (Fe_2O_3), and alumina (Al_2O_3), which are united to form the numerator of the ratio; and the basic compounds, quicklime (CaO) and magnesia (MgO), which are united to form the denominator. The positions as numerator and denominator might, of course, be interchanged, but common practice has established the position as given. Then, for a perfect cement,

$$\frac{SiO_2 + Al_2O_3 + Fe_2O_3}{CaO + MgO} = \text{unity (approximately)},$$

in which ratio the numerical values of the compounds, based upon CaO as a unit (and increased by the proper factor) are to be inserted. Arranged thus, the ratio becomes

$$\frac{2.8 \times 9\% \ SiO_2 + (1.1 \times \% \ Al_2O_3) + 0.7\% \ Fe_2O_3}{\% \ CaO + (1.4 \times \% \ MgO)} = \text{unity.}$$

Derivation of Factors in Index Ratio.—To understand the derivation of the factors in the index ratio (which depend upon the ratios of the molecular weights of the various compounds to that of lime) it will be necessary to explain briefly the calculation of chemical terms, and for this purpose refer to the table of atomic or combining weights of elements, on page 28. Let the compound constituting theoretically pure cement be calculated as an example.

$3 CaO + SiO_2 =$

$3 \times$ (At.wt. Ca + At.wt. O) + (At.wt. Si + At.wt. O)=

$3 \times ($ 39.7 + 12.9) + (28.2 + 12.9)=

166.8 + 60 =226.8

Whence the total chemical weight=226.8. Call this weight 100%. Then the percentage of lime (CaO) is $\dfrac{166.8}{226.8}$=73.4% and the percentage of Silica (SiO_2) is $\dfrac{60.0}{226.8}$=26.6%. For the Index to be unity, then, the percentage of silica present needs to be multiplied by the factors of $\dfrac{73.4}{26.6}$, which equals 2.8. Hence:

Cementation Index.— $\dfrac{2.8 \times \% SiO_2}{\% CaO} = \dfrac{2.8 \times 26.6\%}{73.4\%}$=unity.

Similarly is obtained the factor 1.1 for multiplying into the percentage of Al_2O_3; and the factor 0.7 for the Fe_2O_3.

The factor 1.4 is required to equivalence the MgO with the CaO, and is obtained thus:

CaO $\begin{cases} Ca\ 39.7 \\ O\ \ 15.9 \end{cases}$ =55.6 (1)

MgO $\begin{cases} Mg\ 24.2 \\ O\ \ 15.9 \end{cases}$ =40.1 (2)

Dividing, $\dfrac{55.6}{40.1}$=1.4

Use of Cementation Index.—The practical value of the Cementation Index will appear. Evidently if there is more lime present than can be combined the ratio will be less than one; if there is a deficiency in lime then the ratio will be greater than one. In the case of Portland cement it is aimed to keep this value close to, but above rather than below unity, in order that there may be no free lime or magnesia present. It will be noticed in the table of comparative typical analyses in the preceding section, that the average index for 80 brands of American Portland Cement is 1.09. The Index of some of the Natural cements, too, is on the safe side but one* is the under. For the hydraulic limes it is of course decidedly less than unity, indicating the presence of an excess of free lime.

The Cementation Index is invaluable in calculating the proportions of raw materials required, inasmuch as its value is the same for the uncombined materials as for the finished product. To illustrate this use, let it be required to ascertain the amounts of limestone and clay required of compositions as follows:

	Analysis of Limestone.	Analysis of Clay.
Silica (SiO_2)	2.0	56.4
Alumina (Al_2O_3)	1.0	18.2
Iron (Fe_2O_3)	1.4	3.6
Lime (CaO)	48.4	10.1
Magnesia (MgO)	1.6	1.2
Sulphurous Acid (SO_3)	0.7	2.0
Alkalies (K_2O, Na_2O)	0.3	0.8
Water, Carbon Dioxide	44.6	7.7
	100.0	100.0

Clay

    Silica ....2.8 × 56.4=157.92
    Alumina ....1.1 × 18.2= 20.02
    Iron ....0.7 × 3.6= 2.52
                                 ─────
                    (1) 180.46   180.46

    Lime .... 1 × 10.1= 10.10
    Magnesia ....1.4 × 1.2= 1.68
                                 ─────
                    (2) 11.78   11.78   168.68

                Obtained by subtracting (2) from (1)

*The Akron-Buffalo.

Limestone

Silica ....2.8 × 2 = 5.6
Alumina ....1.1 × 1 = 1.1
Iron ....0.7 × 1.4= 0.98
_____
(3)   7.68   7.68
Lime .... 1 × 48.4= 48.4
Magnesia ....1.4 × 1.6= 2.24
_____
(4)   20.64   50.64   42.96
(Obtained by subtracting 3 from 4)

It will be noticed that the method is to obtain the values of the numerator and denominator of the **Cementation Index,** for each material separately and then to subtract the smaller value from the greater. The next step is to divide the net values, and the resulting dividend is the number of parts of the latter needed to one of the former. Thus, in the example in hand, dividing the net values

$$\frac{168.68}{42.96} = 3.9$$ i. e., 3.9 parts limestone are needed to one of clay.

However, as all lime, theoretically required, can hardly be combined, it is usual to allow 10% reduction. Hence 3.9 less 10% gives 3.5 parts of limestone to be used to one of clay.

Such a calculation as this is necessary in all cases when the original proportions of raw materials are being arranged, or a change is made in the source of supply; and if the rock varies much in composition as the working advances, then this calculation must be frequently repeated. Ordinarily, after a suitable blend is once obtained, the raw material may be more simply compared thereafter by establishing a ratio between the carbonates and insoluble material, which may be readily determined. With such a ratio, the chemist can keep a fairly close watch on the quality of the output. The insoluble material comprises the three oxides used in the numerator of the Index ratio; the carbonates are the soluble material. Thus a comparison of these two—the soluble and insoluble,—forms a fair parallel to the results obtained from the actual indices themselves, to calculate which requires a complete chemical analysis of the materials. The ratio of the soluble to the insoluble matter is usually referred to as the **Lime-Silica Ratio.** For the ideal composition, its value would be approximately three **(3).**

CHAPTER V.

THE MANUFACTURE OF PORTLAND CEMENT.

Classification.—The purpose of this chapter is to familiarize one with the general details of the process of manufacture of Portland cement. It is not intended to cover the ground thoroughly enough to prepare one to design or operate a cement plant, but to impart such knowledge of the subject as is vital to a broad, comprehensive grasp of the field. One may never do more than apply cement in the simplest manner, but it goes without gainsaying that the better one is posted on the steps that go to prepare for his use the material at the basis of his art, the better qualified he will be to use the material intelligently and to keep abreast of progress.

For convenience, the process of manufacture of Portland cement may be considered under 6 headings: Quarrying of raw materials; crushing and drying of materials; grinding and mixing for calcination; the operation of calcination or clinkering; the grinding of the cement clinker; and the packing into bags or barrels ready for marketing.

On page twenty-two were given the various classes of raw material according to which the industry in the United States may be classified. For convenience, these will be repeated:

(1) Cement rock and clay (or shale).
(2) Hard limestone and clay (or shale).
(3) Soft limestone and clay (or shale).
(4) Marl and clay (or shale).
(5) Alkali waste and clay (or shale).
(6) Blast furnace slag and pure limestone.

If these were to be given in order of importance, the ranking would be: No. 1, No. 6, No. 4, No. 3, No. 5. The first is by far of greater importance at present, as it is the combination used in the famous Lehigh region in which Portland cement was first manufactured in this country and which at the present time continues to produce over half the total output of the entire country. Of course, as the manufacture of cement becomes more widely and uniformly distributed over the continent—as it must, for the supply of raw materials is at hand most everywhere—this proportion will diminish. The second class is of growing importance, the third is not and never has been popular, while the fourth is being eliminated by economic law, since it cannot compete in cost with the hard rock producers. The fifth class, which utilizes the by-product of caustic soda and ammonia production, is represented by only one or two plants in this country and, forever limited as it is by its source of supply, will never be an important producer. This by-product is very often almost pure calcium carbonate (limestone)—whence its availability for the manufacture of cement. The drawback to using material of this kind is the likelihood of the presence in the by-product of more or less sulphuric matter which is difficult to remove and which, if not practically eliminated, is injurious to the cement. The sixth class—utilizing blast furnace slag, a by-product of the iron industry—is of increasing importance. The United States Steel Corporation, operating under separate companies, is at the present time capable of producing in excess of 20,000 barrels of high grade, true Portand cement per diem, and it is likely that from time to time this output will be largely augmented.

For the purpose of comparing as to manufacture, these six classes may conveniently be reduced to two, grouped according to the condition of the raw material, whether wet or dry. The wet process differs from the dry in the stages preliminary to the burning, and partly through the burning stage also. In the wet process the slurry, as the creamy mixture of marl and clay is known, is fed into the kiln without preliminary drying, although the amount of moisture present may be and is generally considerably reduced by compression in powerful hydraulic presses. But in any event the operation in the kiln is protracted and requires much more fuel. The additional expense of this is largely offset by the reduced cost of preparing the raw materials: both crushing and coarse grinding are eliminated and fine grinding very much facilitated. It is estimated that less than half as much power is required to pulverize wet material as dry. Wet producers, however, are at a considerable disadvantage in northern climes, as the marl beds freeze up with the advent of cold weather, necessitating the shut-down of the mill until spring.

The details of the Dry Process will be first described:

Quarrying of raw materials.—Clayey limestones are easily quarried directly by steamshovel—hard limestones more difficultly, requiring almost always considerable blasting. The use of the steam shovel is a growing custom in the Lehigh district, where the cement rock outcrops in steeply inclined layers, peeling off readily when undercut. To facilitate this method of quarrying, steam shovels of a special type have recently been put on the market. These shovels are said to be capable of excavating shale or hard cement rock at the rate of 5,000 to 6,000 cubic yards a day, and require no blasting. Such excavators are in use at the plants of the Atlas, Edison, Sandusky and other large cement companies. Biting off chunks of solid rock 5 cubic yards in mass, these shovels are able to meet almost any demand likely to be made upon them. Only three men are required to operate a shovel; two engineers, one for the lifting, the other for the revolving mechanism, and a fireman.

In many of the plants the old type of shovel is still in use— a shovel which is not intended to attack the rock directly but only handles it after cast down by blasting and reduced to chunks, either by sledging or secondary blasting, of a convenient size. The shovel is then used for rapid loading into conveyors for transportation to the crusher. The demand for raw materials is so great when a mill is operating at capacity that a complete mechanical equipment is as essential at this end of the process as at the other. As the efficient and successful operation of the mill depends upon the uninterrupted sequence of every detail of the process, from native rock to finished product, the general adoption of the new style excavators is likely.

In opening up a quarry of cement rock an extensive series of preliminary borings must be made in order to check up the geological data and the preliminary surveys that have been made to determine the location and extent of the outcrop. As the composition of the rock is liable to change with the depth or thickness of the layer sounded, it is necessary to sink the borings to the bottom of the layer, and from time to time, as the working advances, to continue these borings. It is upon the results of tests made on these samplings, that the proportioning of the materials is made. Obviously, any considerable variation in the nature of the rock—say, for example, that its percentage of clay suddenly doubled—would, if not detected, destroy the value of the cement, for a true Portland is an exact—not nearly exact—combination, and must be as carefully compounded as the pharmacist's prescription.

Mills have been known to get in trouble in exactly this way. On account of failure to keep careful check on the composition of the rock, as the quarry working advanced, an alteration in the composition, unnoticeable to the operatives perhaps, got by and for several days cement of a very inferior grade was being produced. If a shipment of such cement were to be accepted on the prestige of the brand,—as it is often, though unwisely, done,—and were it to be used in any important work, trouble—even failure—might result. This is an important point for the user to keep in mind. It emphasizes the need for always carefully testing all cement. Of course, old and reliable mills are not likey to let anything like this slip by them—but, as the old saying has it, "accidents sometimes happen in the best regulated of establishments," so that as long as the manufacturer of this or any other article is dependent in the slightest upon the human element, the product will never be "fool-proof" and will require careful checking up or testing at all important stages. Better far to be on the safe side — ALWAYS TEST CEMENT PROPERLY BEFORE USING.

The other ingredient—clay, either in the form of earth, shale, or slate (slates have not been used to any extent but are equally available) in the meantime is also being excavated. Shale is may be attacked readily with steam shovel without the preliminary

The other ingredient—clay, either in the form of earth, shale, or slate (slates have not been used to any extent but are equally available) in the meantime is also being excavated. Shale is handled the same as the cement rock, although, there being relatively much less of it to secure, the apparatus required is correspondingly less extensive. Shale rock is comparatively soft and may be attacked readily with steam shovel without the preliminary of blasting. Clay occurs generally in valleys, along the margin of streams, being a deposit of fine particles eroded from the rocky hills above. It is in layers several feet thick, often as deep as 10 or 15 feet and again only in a thin surface skim. An excavation of a clay deposit is known as a borrow pit. The excavating or borrowing, after the surface soil has been stripped off, may proceed with the use of steam shovels or wheel scrapers, the material being loaded into skips or cars.

Some plants, not conveniently situated to a suitable deposit of clay or shale, find it to their advantage to purchase their supply from specialty companies, who arrange to deliver in any desired quantity. This of course means a slight increase in the cost.

Crushing and Drying of Materials. — The raw materials —the quarried rock and the borrowed clay—are transported to the site of the plant variously, by tram-car, cableway, automatic conveyor, or even by wagon. The use of **electrically** driven apparatus for this purpose is being introduced and is meeting with increasing favor, especially among those plants that employ the electric drive throughout the process of manufacture. **Steam, compressed air,** and **gas** are among other favorite sources of power for mechanical conveyance. **Animal power** is no longer efficient in connection with this highly individualized.industry. Some plants are so fortunately located with reference to the supply of raw materials that they are enabled to avail themselves of **gravity**—the cheapest known source of power—to get the material from the quarry to the mill, requiring only power-haulage to return the empty skips or cars. It might be observed that this is a vital consideration in determining the location and often offers an advantage that may determine the financial success of the mills.

Reaching the mill, the clay—if wet. is first put through a rotary dryer and then is introduced into a disintegrater and reduced to powder. It is then ready to be mingled with the pulverized rock for the finer grinding and intimate mixing necessary to prepare it for the kilns.

The cement rock on reaching the mill is fed into a rock breaker, generally of the gyratory type. This is so placed in regard to the receiving incline as to take the chunks of rock directly from the dump without further handling. The rock issues from the breaker in sizes up to about a 2-inch diameter—about right for road metal —and is elevated by bucket hoists to be screened and binned. Usually the supply of rock and the capacity of the breaker is sufficient to allow the use of the larger sizes for concrete and road metal, the screenings and dust being sufficient to keep the cement plant operating.

In case the excavation is done by the huge steam shovels, as afore noted, which often deliver chunks of solid rock up to ten tons each, it becomes necessary either to introduce an intermediate breaking operation, or to have a special type of breaker capable of handling directly the largest chunks. To meet this need, there have been marketed immense **gyratory crushers** capable of handling the largest pieces practicable to take from the quarry, without the need of sledging or secondary blasting. Such a crusher is shown in Fig. 1. Some idea of its size may be had by comparison with the figure of the man standing on its base. While this particular type of crusher was not primarily intended for use in con-

FIG. 1.   A GYRATORY CRUSHER (Allis-Chalmers Make).

nection with a cement plant, it is nevertheless meeting with favor for this purpose. Its capacity is practically unlimited, for where used it has been found impossible to bring rock fast enough from the quarry to choke it.

As an interesting development, it should be noted that the demand for crushed limestone, not alone for cement manufacture but for concrete construction, track ballast, and road metal, has in some sections become so enormous that its quarrying and crushing have been undertaken as a separate industry. Usually the quantity of stone available is so vast that its use for these other purposes in no way jeopardizes the cement manufacturer.

The principle of the **gyratory crusher** is an interesting one. The mouth of the machine is an inverted cone; the breaking head is also conical but is upright, and is suspended at the upper end from a universal joint supported by a spider frame resting on the side of the machine. The shaft, at the upper end of which the breaking head is set, revolves through an annularly slotted plate at its lower end. This imparts a combined rotary and swinging motion to the head, which makes it act like a roll against an annular die, always crowding the material ahead of it while allowing freedom to that trailing so that it can drop through as the opening between the head and the concave mouth widens. As soon as the material is fine enough—and the machine admits of ready adjustment to the size required—it drops through to the chamber below and is discharged through a spout into conveyors and thence elevated to bins for storage. There is thus a continuity of action throughout and a gradual reduction of material to the desired fineness—a combination that is ideal.

The Edison Portland Cement Plant, Stewartsville, N. J., among other unique features, includes a breaker of decidedly original yet simple design. In brief, it is merely a series of three parallel rollers, corrugated faced, superimposed one above another, the upper pair of which are set some distance apart so as to be able to bite the largest fragments of rock liable to come from the quarry; and each lower set at progressively closer spacing so as to catch and crush the chunks passing the rollers above. The last or lowest pair of rollers are set close enough together to crush finally to the desired fineness. This breaker is said to be very efficient and its capacity enormous.

Crushing of the Coal. — The question of fuel is a very important one. Some plants use oil or natural gas, which requires no preliminary treatment, being fed into the kilns at the lower or discharge end by compressed air or steam-jet injectors. Most plants however, use coal—either anthracite or bituminous, generally the latter. This requires to be crushed, dried, and pulverized, just as thoroughly as the other constituents. Some idea of the part played by the coal and the extent of the apparatus necessary for preparing it may be grasped from the statement that for every barrel of cement produced — weighing about 380 pounds — 100 pounds of pulverized coal is required. This all disappears in the operation of clinkering, its value appearing in the chemical and physical combination incidental to calcination.

Ordinarily, the only apparatus required for crushing the coal is a pair of iron rolls, either tooth or corrugated faced. This is a simple apparatus and requires no special description. From the rolls the granulated coal is passed through another disintegrating apparatus (grinding machinery used for reducing the stone is equally applicable here) whence it enters the dryers. As the dryer used for the coal is similar to those used for the other material, a description of it is reserved until later. From the dryer, the granulated coal is passed through a fine grinder, usually a tube-mill, and reduced to at least 90% on a 100 mesh screen. The finer the coal, the better its action in the kiln and the greater efficiency realized from it. The powdered coal is then conveyed to bins, there to be drawn upon as needed for injection into the kiln.

The Drying of Materials. — The raw materials—limestone, clay, and coal—as they come from the quarry, pit, or mine, always contain more or less moisture. This must be removed by evaporation prior to the operation of pulverizing, if that operation is to be successfully and economically performed, inasmuch as the moisture freed by grinding would tend to ball up the mass, clog the machinery, and hinder the reduction.

The most efficient and common dryer in use is simply a small rotary kiln. Indeed, the rotary dryer is the only one adapted to a continuously operating plant, such as most up-to-date plants are. It is a steel cylinder, four to seven feet in diameter and 30 to 80 feet long, set at a slight inclination to the horizontal and revolving at the rate of two to five times per minute. The fire box is stationary, and may either be set at one end or may encase the entire cylinder. In the latter case, the entire cylinder between carrying rings, which are outside the fire box, is divided by air spaces into longitudinal compartments, each one a complete separate chamber. Thus, on all sides of each compartment, through the spaces between the dividing plates, the heat gases have free play. This is said to provide a very efficient drying action,—one both rapid and economical. It is adapted to drying either crushed rock, ground clay, or fine coal. Such a dryer (according to the maker, the C. O. Bartlett & Snow Co., Cleveland), may be relied upon to evaporate six pounds of water per pound of coal when used to dry coal. Coal as it comes from the mine contains about 12% water. Thus, allowing one pound fuel coal for each six pounds evaporation, only 40 pounds coal per ton per hour would be required. An illustration of such a rotary dryer is given in Fig. 2. This is only one of a number of good dryers in use.

FIG. 2—FOUR-COMPARTMENT,
DIRECT-HEAT, ROTARY
DRYER.

The material to be dryed is fed in at the upper end, assisted forward by gravity and is repeatedly upheaved by mechanical lifters affixed to the inside of the dryer. At the discharge end it enters a dust settling chamber, whence it is taken up by automatic conveyors and delivered to the grinders.

The heated waste gases from the rotary kilns are often utilized to supply the dryers; but it is not always convenient to apply them on account of the distance ordinarily required between the two.

Grinding and Mixing.—This stage in the process is usually divided into two parts—coarse grinding and fine grinding the former being conveniently known as **granulating** and the latter as **pulverizing.** In the former operation the materials are still separate; in the latter they are for the first time brought together and the operation of final reduction and intimate mixing accomplished simultaneously.

It should be noted that there are grinding mills, among which may be mentioned the Kent, Griffin, and Fuller-Lehigh, that are capable of reducing at one operation to the required degrees of fineness either the raw material as it comes from the crusher or the cement clinker. The Raymond Bros. Impact Pulverizer Company, of Chicago, also have a mill adapted for the complete operation which has especially met with favor for the grinding of the coal. It is noteworthy as combining an air separator for removing the fine material as fast as it accumulates, and an apparatus for insuring dustlessness of operation. The Aero Pulverizer Co., of New York, also make a pulverizer of similar qualification. An equipment of light machines is in use for preparing the coal in the Pittsburgh plant of the Universal Portland Cement Company. While this combined type of grinding mill is meeting with favor for preparing the coal, it seems to be the generally favored practice to conduct the reduction of the other materials separately. The chief reason for this is that the necessary intimate mixing of the raw materials, the clay and the limestone, can be much more thoroughly effected when they are not brought together until already in a finely divided form. The operations of mixing and final reduction then go on together.

The **first stage** of the reduction is usually accomplished in a **Ball Mill,** and the final in a **Tube Mill.** The **first** is known as a **coarse grinder** and the **latter** as a **fine grinder.** For the coarse grinding the Kent, Griffin, and Fuller-Lehigh Mills are also much used. A more recent coarse grinder, for use at the raw material end only (the other mills are applicable at either or both ends of the process, for grinding clinker into cement as well as grinding the raw material) is the Williams Universal Grinder.* It is claimed for this machine that it will take 1½ inch dry raw material and reduce same to 95% fineness on a 20-mesh sieve, giving an excellent tube mill feed. The capacity of this machine is said to range from 700 to 800 barrels per day of 24 hours, the horse power required from 45 to 50. The cost of this operation is said to be hereby reduced to half a cent or less per barrel. Machines of this type have been installed in the new plant of the Sandusky Portland Cement Co., at Dixon, Illinois. They are said to be very economical and admit of ready repair.

The old style **Ball Mill** is simply a steel drum, with a diameter twice or more its depth. It revolves upon a horizontal axis at a

---

*(Made by the Williams Patent Crusher & Pulverizer Co., of Chicago.)

rate of 21 to 27 times per minute. The grinding is accomplished by means of cast steel balls, with which the cylinder is partly filled. Feed is through a hollow trunnion and discharge through a perforated shell, concentric encompassing screens returning the tailings as fast as collected through the perforations in the cylinder to be reacted upon. This type is being replaced by a modified mill (see Fig. 3),

FIG. 3—SECTIONAL PERSPECTIVE VIEW OF BALL-TUBE MILL.

which is no longer a drum but a thick cylinder. It is frequently known as the **Ball-Tube Mill**. The interior of the Ball-Tube Mill is lined with a series of perforated overlapping or stepped-up plates, which serve to lift the balls and cast them on the material, crushing it to powder as the mill revolves. The steel balls, which are forged from high carbon metal, range in size from 3 to 5 inches, and fill the chamber about one-third. A perforated partition at the lower or discharge end allows the ground material to escape while retaining the balls. The feed, just as it comes from the crusher, enters the Ball-Tube Mill through a hollow trunnion, and leaves similarly, being reduced in transit to a fine grit, 95% of which will pass a No. 16 or 20 seive. The mill with its charge of balls and material is very heavy and requires considerable power to start it. A low speed electric motor, direct connected to the counter shaft, with its ability to stand a heavy overload, is thus especially adapted for driving these mills.

The clay requires no crushing or coarse grinding as a rule: but simply requires to be passed through a set of disintegrating rolls, which mash it to powder. It is then ready to be dried and passed on to be fed into the final grinders simultaneously with the granulated limestone. The clay ingredient, if in the form of shale rock, requires not only to be passed through disintegrating rolls as before but to be put through the regular sequence of grinding.

Interposed between the operations of coarse and fine grinding of the raw, materials is the **weighing apparatus,** which automatically receives the powdered clay from the disintegrators or the granulated shale from the Ball Mills, and the granulated limestone from the Ball Mills, weighs them in exact proportions, and delivers them simultaneously to the Tube-Mill Feed hoppers. The weighing pans or hoppers are in pairs so adjusted, according to the amount of each material needed as previously determined, that the one can not deliver its charges until the other has its correct amount of material.

(Scales for this purpose are made by the Power & Mining Machinery Co., Milwaukee. the Automatic Weighing Machine Co., Newark, N. J., and the Richardson Co., New York.)

For the **final reduction,** either of raw material or of clinker, the **Tube-Mill** (Fig. 4) is almost the universally accepted medium, and may therefore be considered standard. A Tube-Mill is a steel cylinder, usually from 20 to 22 feet long, and 5 to 6 feet in diameter. The larger the diameter the more efficient the grinding. The mill is about two-thirds filled with flint pebbles, the most of which are about 1½ inches in diameter. The pebbles are retained by means of a coarse screen fitted in the cylinder near the discharge end. The shell is lined with some very hard material, either silex stone (a flint stone imported from Europe), ironite (a dense, close-grained stone native to U. S.), porcelain or chilled iron. In the wet process it may be lined with wooden blocks. It is the action of the pebbles against one another and against the lining that accomplishes the pulverizing. Obviously the grinding surface is very great, a fact that permits of a very slow speed of rotation, 22 to 27 r. p. m. according to the size. Hence arises the high efficiency of this machine and its general preference for all fine grinding, either of raw materials or cement clinker. The granulated raw materials, nicely proportioned by the automatic scales, enter the mill through a hollow trunnion at one end and are delivered similarly at the other end reduced to the required fineness, and perfectly mixed.

By simply regulating the feed, any desired degree of fineness almost may be secured. The determining point as to the degree of fineness is where the increased cost of grinding is no longer justified by the increased strength realized in the product. If the standard should be raised, then the cement must be ground finer to meet the requirements and the consumer must be satisfied to pay a better price for a better article. It should, however, be noted that the consensus of opinion among experts is that the limit has already been practically reached. Hence much finer grinding is hardly to be expected. Thence they are taken up by automatic conveyors and delivered to storage bins, to be drawn upon to feed the kilns as required. The Tube-Mill is about as **perfect a fine grinder** as could be devised, and being simple in construction and operation is also very economical.

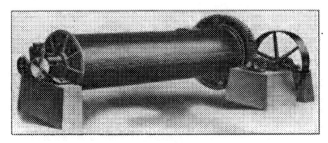

FIG. 4—DRY GRINDING TUBE MILL DRIVEN BY SPUR-GEAR
FROM DISCHARGE END.

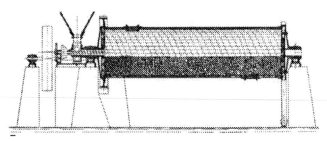

FIG. 4a—LONGITUDINAL SECTION OF A TUBE MILL.

A 5 x 22 Tube-Mill, which is a favorite size, will pulverize—if properly fed—from 14 to 20 barrels per hour, depending on the hardness, age, and fineness of the clinker. It will require from 70 to 80 horse power to operate it when once started, and about 125 momentarily when starting. Illustrations, showing general appearance and section, are given in Figs. 4 and 4a.

The **three special grinding mills** above mentioned, the **Kent, Griffin,** and **Fuller-Lehigh,** are so frequently used, as **coarse grinders** chiefly, at both ends of the process, that a separate description of each is included. To facilitate the description, cuts of these machines are also given.

The grinding principle of the **Kent mill** is that of a roll exerting a constant pressure against an annular die. The material to be ground is caught between the roll and the ring and is thereby reduced to powder. The Kent mill (see Fig. 5) embodies three rolls, identical in every respect, and each held against the periphery at a constant pressure by means of powerful springs re-

---

(The Bonnot Co. of Canton, O., the Power & Mining Machinery Co., Milwaukee, the Allis-Chalmers Co., Milwaukee, and the F. L. Smidth Co., of New York, are the principal makers of these mills.)

acting against the frame of the machine with a pressure adjustable up to 20,000 lbs. Only the upper roll is revolved by power. Pressing against the annular die which is free to turn, it causes it to revolve and this in turn being in contact with

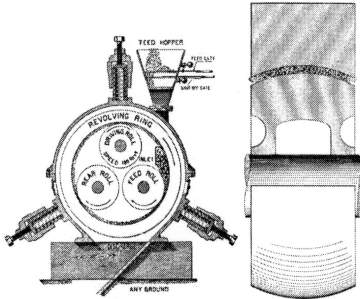

FIG. 5—INTERIOR OF KENT MILL.

FIG. 5a—SECTION OF ROLL AND RING OF KENT MILL.

each of the other rolls transmits its motion to them. The face of the rolls is slightly convex and that of the ring correspondingly concave, so that the coarser particles are retained in the hollow, while the finer ones are gradually squeezed out to the edge of the ring where they drop off and are delivered through a spout. After the motion is once begun the material is introduced and the centrifugal force developed causes it to follow around the periphery in a smooth thin layer, about an 1 inch thick, passing in succession beneath the rolls. The rock is thus caused to crush itself very largely. In fact it is claimed by the makers of the machine that fully 90% of the rock is reduced by abraiding on itself, and there is no rubbing of parts on the rock, so that the wear and tear on the parts is reduced to a minimum—an important consideration, as the repair item in cement mills is very large. A modification of the Kent mill, which is simply a larger and more solid machine, has been put on the market recently. This new type is called the "Maxecon," because it is held to combine maximum efficiency with maximum economy. A battery of these is in use at the Chicago works of the Universal Portland Cement Company. The makers are the Kent Mill Company, New York City. A horse power of 25 is all that is required to operate a "Maxecon" Pulverizer.

The Griffin Mill* operates on a rather unique principle. It, too, utilizes centrifugal force, not to distribute the material in a thin layer against a ring there to be acted upon by the rolls, but to hold the roll itself against an annular die. This is accomplished (see Fig. 6) by suspending the upper end of the shaft, at the

FIG. 6—COMPOSITE FRAME MILL,
OPENED UP.

lower extremity of which the roll is hung, from a universal socket. This allows the roll, as the revolution is begun, to swing out by centrifugal force until it bears hard against the encompassing ring, and the material being caught there between is pulverized. A strong point about this mill is the fact that all bearings are outside the grinding chamber, which in a process entailing so much dust in the grinding parts is most desirable. The latest Griffin (Fig. 7) contains three rolls instead of one, but the general principle is the same. The three rolls are suspended

*(Made by the Bradley Pulverizer Company, of Boston.)

from a revolving plate above and are set in sockets which allows them while turning freely to swing out as before. More thorough grinding, greater unit capacity, and greater efficiency are claimed for this new machine.

FIG. 7—GRIFFIN MILL, 3-ROLL MILL, OPENED UP.

The Huntington Mill is another type of coarse grinder utilizing the centrifugal principle. It somewhat resembles the Griffin Mill in make-up and action, especially the new Three Roll Griffin. It is said to be very efficient. It is in use throughout and only in the plants of the Atlas Portland Cement Company, who own and control the patents.

The Fuller-Lehigh Pulverizer* is one of the best known grinders. In principle it somewhat resembles the Kent Mill, in that the grinding is accomplished by the impact of revolving steel against an annular die; but in this case the crushing is accomplished by use of steel balls which are propelled around the chamber and urged against the die by means of arms attached to a central revolving shaft. There are four balls. Above the balls are stationed two fans superimposed: the lower draws the pulverized material up out of the grinding chamber and the upper fan floats it outward to the discharge. There are two sizes of this mill. The 33" mill uses balls 9½" in diam., propelled at the rate of 200 to 210 revolutions per minute; the 42" mill balls of 12" diam., propelled at a speed of from 150 to 160 revolutions per

*Made by Lehigh Car Wheel and Axle Works, Fullerton, Pa.

minute. A battery of 16 of these are in use in the raw material department of the German-American Portland Cement Company, Alsen, N. Y.

There is another type of coarse grinder, known as the "Kominuter"* (see Fig. 8). This is a modified type of ball mill, in which

FIG. 8—KOMINUTER, SHOWING AUTOMATIC TABLE-FEED AND RETURN PIPES FOR TAILINGS.

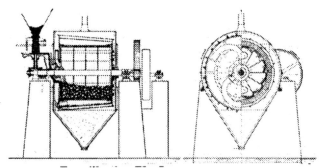

FIG. 8a—LONGITUDINAL AND TRANSVERSE SECTION OF THE "KOMINUTER."

*F. L. Smidth & Co., New York.

it is claimed the disadvantages of that mill have been eliminated. In general make-up is closely resembles the Ball-Tube mill previously described. The great point of difference and advantage over the old type ball mill is that it embodies a perforated enclosing shell which allows the fine material to drop through to the encompass-

FIG. 9—ROTARY KILN.

ing screen and be separated and removed as quickly as possible. The screen catches the particles that are still too large and continually returns them by means of specially curved pipe chutes attached to one end of the drum, to the interior to be reacted upon. The mill is encased in a dust-proof housing. Its design

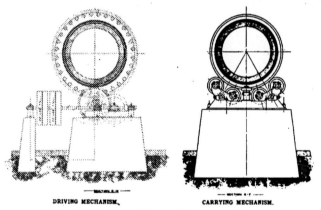

DRIVING MECHANISM,   CARRYING MECHANISM.

FIG. 10—ROTARY KILNS—SECTIONAL VIEWS.

also permits of a direct drive. It is claimed that a "Kominuter" gives 65 to 85 per cent more output per horse power than a ball mill. Kominuters are being used in the raw material department of all the Lehigh Portland Cement Co.'s plants.

Calcination.—The operation of calcination or clinkering—by which the correctly proportioned and intimately mixed and powdered limestone and clay (or shale) are freed of their combined water and carbon dioxide gas, and the peculiar formation of substances which constitute Portland cement, is effected—is accomplished now almost entirely in a rotary kiln. This is a long steel cylinder (see Figs. 9, 10, 11) set at a slight inclination to the horizontal ($\frac{3}{4}$ to 1" to the foot) and revolved at the rate of about once per minute. The kilns of smallest diameter may be revolved 2 or 3 times a minute; the largest only $\frac{1}{2}$ to $\frac{1}{3}$ revolution per minute. The normal speed of revolution of a 150 foot kiln, diam. 8 feet, is $\frac{2}{3}$ rev. a minute. (See table at the end of this chapter.) At the upper end of the kiln the ground raw materials are fed in,

FIG. 11—CARRYING AND THRUST MECHANISM.

are carried forward and tumbled over and over by the slant and revolution of the kiln, and as they advance are met and reduced by the hot gases from the burning fuel which is being fed in at the lower end. The contents of the kiln emerge at the lower end in small particles, at white heat, ranging in size from walnut to pea. Upon cooling they assume a semi-vitreous appearance, dark gray in color. The transit of the material during which it is converted into Portland cement clinker, occupies upwards of an hour. In the 150 foot kilns of the Edison Portland Cement Company's plant, for instance, the material takes about 1½ hours to pass through. The physical and chemical changes that take place are progressive and occur as follows; first the water of crystallization (chemically combined and not removed by the preliminary drying) is expelled; then the carbon dioxide CO_2, combined with the calcium, Ca, as calcium carbonate or limestone is released as a gas, to be carried off with the rest of the gases up the flue; and finally the combination at the state of incipient fusion of the lime and silica as tri-calcic-silicate (3 CaO) (Si O_2), which is pure Portland cement. (See Chapter IV.)

The alumina, iron, and magnesia present likewise unite to form analogous compounds. These minor compounds also act as a flux to facilitate the fusion of lime and silica. (See Chapter IV). Considerable care must be exercised that exactly the proper amount of heat is supplied, as both underburned (partially combined) and overburned (completely vitrified) clinker impairs the quality of the cement. Partly overburned clinker is objectionable because of the resistance offered to grinding, which impairs the effectiveness of the grinding machinery. Completely over-burned, or vitrified clinker, is worthless as cement, being only so much inert matter. The underburned particles, being composed of free lime and other incompletely combined compounds, is a source of real danger to the cement. The effect will be to impair the soundness and tensile strength. Such cement getting into finished work, would cause it to discolor and possibly in the course of time to disintegrate. Hence arises the necessity for careful testing to detect such weaknesses.

Tests made to determine the fitness of a cement include a determination to detect the presence of underburned particles. These will show up as a bright yellow jelly when a portion of the cement is treated with muriatic acid in a test tube. (See laboratory section, testing of cement.) If all of the silica is not combined it will appear as so much insoluble fine grit, which is harmless to the cement, simply constituting so much more inert matter to reduce the sand carrying capacity.

The temperature required in the kiln, to effect the reduction and efficient re-combination of the compounds, varies according to the nature of the material from 2500 to 2800 degrees Fahrenheit. The influence of the constituents on the degree of burning necessary to effect the complete change was dwelt upon in Chapter III. A review of these paragraphs at this point is desirable.

The temperature is of course regulated by the draft. The exact degree is a matter of experiment, and must be adjusted from time to time to suit changing conditions. When once adjusted so as to produce the proper clinker, the attendant maintains this temperature by observing the color of the flame, through colored glasses.

If the clinker were to be ground up and used just as it emerges from the kiln it would take an almost immediate set: to regulate this a quantity of gypsum, about 2%, is added to the clinker before grinding; or of ground plaster, about 1%, to the ground cement. It is preferable to add this set-retarder prior to the grinding if possible, as it then can be more intimately incorporated. Mills have been known, either through lack of an automatic gypsum feed or failure of attendant to keep same supplied, or perhaps through an inferiority in a fresh supply of gypsum, to deliver cement, supposedly alright, which when tested took an immediate set. This is an important fact for the user to know, as it may enable him one day to explain the unusual behavior of a lot of cement when the mill is unable to do so. Such mistakes are more liable to happen when the gypsum is added to the ground cement than when it is previously incorporated—a fact which constitutes an additional reason for its addition to the clinker.

The uniformity of the product turned out by the rotary kiln, when once nicely adjusted, is marvelous. Formerly, with the dome kiln, it was almost impossible to secure a uniform product, and the sorting out of imperfectly combined particles to be returned to the kiln and of overburned particles to be cast aside combined to increase the cost of the process. The introduction of the rotary kiln and almost immediate and universal adoption, in 1888 to '90, was revolutionary in its effect, and served to put American Portland cement to the front at a jump and to give an impetus to the industry well nigh incalculable. The increase in the consumption of cement from less than 400,000 barrels in 1890 to 55,000,000 (approx.) in 1908, an increase of over 150 fold, tells the story. The cost of production was hereby cut in half and the standard almost doubled.

The early rotaries were pigmies compared to those now being adopted. A kiln 5 feet in diam. and 50 feet long was then the limit. Now engineers favor kilns 7 to 8 feet in diam. and upwards of 100 feet long. The kilns installed in the plant of the Edison Company are 8 feet in diam. and 150 feet long, and have an output per kiln per day of 24 hours, of 800 to 1,000 barrels. Kilns 130 feet long are now common. The Atlas Portland Cement Co. are reported to be building kilns 10 feet in diam. and 240 feet long. Whether they will be successful or not remains to be seen.

The kiln is in two sections: the exterior or body is of heavy steel plates, heavily reinforced at all important points and well riveted up, for in a long kiln when charged and revolving the bending strains are enormous and require the most substantial construction; the interior, which comes in contact abraisively with the highly heated materials and gases, is composed of a layer of the hardest and most refractory brick possible to obtain. These are simply laid up inside of the shell and held in place until the arch is completed by means of angle shelves affixed to the shell.

The kiln is fitted with heavy, cast iron riding rings, encompassed by steel bands or tires shrunk on. These rings—generally two in number—are placed at about the quarter points of the kiln so as to equalize the bending strains as much as possible. They rest in and revolve upon a nest of rollers, usually four in number, two on either side. The roller carriage is also provided with an auxiliary twin-faced roll, rotating in a horizontal plane, which bears on either side of the riding ring, thus taking up any tendency of the kiln to slide either way. Usually it is necessary to have this thrust mechanism, as it is technically known, on the upper carriage only.

Somewhere between the upper riding carriage and the middle of the kiln, usually quite close to the said carriage, the driving mechanism is located. It consists of a train of gears, actuated by belt and pulley or by direct connected motor, and engaging the kiln cylinder through a heavy cast-steel gear-toothed ring, which resembles the riding ring in manner of attachment. It is so arranged that two or more speeds are possible—quite slow for start-

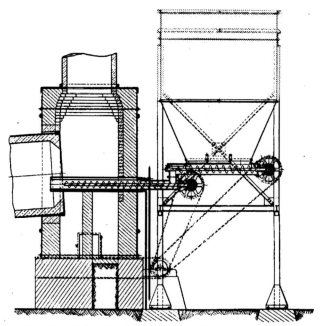

FIG. 12—RAW MATERIAL STORAGE BIN, WITH FEEDER, DRIVEN FROM KILN COUNTERSHAFT.

ing and faster as the operation gets well under way. By regulating the speed of revolution together with the rate of feed the output and quality of the clinker is controllable.

The upper end, through which the kiln is charged, is sometimes tapered so as to hasten the forward travel of the material away from the feed and to equalize the heating effect of the fuel. In the dry process this coning may be omitted, but in the wet process, where the slurry is fed directly into the kiln, it is practically indispensable. This upper end terminates in a brick flue chamber, which serves the double purpose of carrying a stack to remove the products of combustion and of providing support for the charging tube. The bins, in which the prepared raw material as it comes from the final grinder is conveniently stored, are located just beyond this flue chamber, and may be discharged into the kiln by means of screw pipe conveyors (see Fig. 12). A common form of screw conveyor is in two sections—one reaching from the discharge orifice of the bin to the flue chamber; the other extending through the flue into the orifice of the kiln. The latter is water jacketed to protect it from the intense heat of the gases. Its mechanism may be actuated by a connection to the countershaft of kiln driving pulley, so that the two act in harmony, giving an automatic feed. The upper section of this screw conveyor, ordinarily an open trough, is actuated independently, and has its flow controlled by a slotted, graduated crank, by keeping a record of whose positions a comparative record of feeds may be secured.

The lower or discharge end of the kiln is fitted with a firing hood, which rests on a four-wheeled truck and rides on rails. It may thus be removed whenever it is desired to get at the interior of the kiln for cleaning or repairs. It is composed of a heavy cast ring into which a heading of fire brick is placed. An opening is provided near the top for the admission of the fuel, and at the bottom on the under side of the hood another opening for the clinker to emerge from the kiln. The fuel, if powdered coal, is fed in continuously by means of an air blast; if oil, by means of a steam jet; if gas, by its own pressure. It is ignited as it emerges from the "injector," or "burner," which extends through the orifice in the hood.

Cooling.—The clinker as it emerges from the kiln is at a white heat, so that before passing on to the grinders it must be cooled. For this purpose rotary coolers are provided, similar to the rotary dryers, with the difference that cool air instead of hot air is directed against the revolving cylinder. If the coolers are located out of doors no artificial air supply may be required.

In the plants of the Atlas Portland Cement Co. use is made of the heat in the fresh clinker to warm the blast of air that is used to inject the coal and force the combustion. This is accomplished by passing the air supply through the rotary cooler into which the clinker is delivered from the kiln. The partially cooled clinker is then passed through a second rotary cooler, when it is ready for the grinders. This make for an economy in the fuel consumption, but it is questionable whether the saving justifies the apparatus necessary.

The cooled clinker may be ground at once; or it may be accumulated for future grinding. In plants where it is usual to keep the plant going at a specified rate independent of the present demand, it is customary to accumulate the clinker during the slack season, so as to have at all times a reserve supply for emergency. Sometimes additional finishing grinders are provided, so that by utilizing this reserve of clinker the output of the mill may for occasion be largely augmented without taxing any other part of the equipment or impairing the quality of the cement.

Grinding of Clinker.—This operation and the apparatus required are practically identical in preparing the raw material, even to the number and size of the mills, unless additional units on the finishing side are provided as previously noted. The clinker is harder to grind than the raw material, but the reduced bulk due to loss by ignition in the kiln more than compensates. It is estimated that fully one-third of the raw materials is lost in ignition. Thus to produce 400 barrels cement requires 600 barrels raw materials.

The tube mill is used almost entirely for the fine grinding, as it not only reduces the material to any desired degree of fineness but eliminates all irregularities, turning out a product of great uniformity. By means of a tube-mill, cement may be turned out fine enough to meet the most rigid practical specifications; for instance, to satisfy the standard specifications of the American Society for Testing Materials, calling for 92 per cent to pass a 100 mesh sieve, and 75 per cent to pass a 200 mesh. Still finer pulverizing is possible with this machine, but hardly practicable. It is said that if the feed is too slow, the material by the excessive pounding is pressed into flakes, so that the product looks mealy. So there is a practical limit to the degree of fineness possible with this machine. It is also possible to grind so fine that the product will mix with water reluctantly, like Lycaepodium.

Storage and Shipment.—The pulverized clinker, which is now cement, as it comes from the tube mills, is automatically elevated into stock bins, where it may be drawn upon for packing into bags or barrels as required. As the cement when freshly ground is still unfit to use, requiring aging or curing usually for a period of at least a month, it is virtually necessary to provide abundant storage for the ground cement. Moreover, it furnishes a reserve of finished product in its most flexible form with which to meet extraordinary demands. In northern latitudes, where during certain months the demand for cement drops to almost nothing, and efficiency requires the plant to be operated, huge stock house storage must be provided. In the south and west where the demand is more uniform this is not required.

Freshly ground cement often exhibits peculiar properties; when subjected to the boiling test it swells and cracks badly. When, however, this same cement after a few weeks' aging is again subjected to the same test, it is unaffected. Just what is the cause of this behavior, which may occasion rejection of the cement, and which is reason sufficient for withholding the freshly ground cement from the trade for a period of perhaps a month, it is hard to say; but it is supposed to be due to the presence in the freshly ground product of infinitely small amounts of free or loosely combined lime, which does not slake freely like ordinary lime, but slowly, requiring perhaps the assistance of heat. It is the presence of these compounds that vitiates the boiling test. During the period of aging or curing these seem to be corrected, probably by combination with the CO_2 of the atmosphere, forming inert carbonates. An abnormal amount of these compounds, indicating gross error in some part of the process of manufacture, will not be thus easily corrected, and will result in an unsound cement, causing rejection. In addition to aging the ground product, manufacturers to the same end practice wetting down the hot clinker, but perhaps the most successful expedient thus far developed is to mingle steam with the clinker in the tube-mill. By the combination of watering the hot clinker and steaming it in fine-grinding, it is claimed that aging becomes unnecessary, and the product is safe to use as delivered from the fine-grinder.

In all events a cement reaching the work and testing unsound, should not therefore be rejected, but held in storage for a few weeks, and the test repeated. Then, if still unsatisfactory, it may assuredly be rejected. Mills with a prestige, or seeking one, are usually very careful in this respect, but with the best mills the stock house may run so low in periods of rush that they will take a chance on fresh material. Then it is that the user needs to be on the lookout, accepting no cement on a basis of prestige, or representations of sale agents.

It has also been pointed out that this trouble might be met by aerating the freshly ground cement. However, this does not seem convenient to do at present. Some plants realize considerable aerating by the method of removing the tube mill delivery to the stock house. At the Universal Portland Cement Works, Buffington, Indiana, the stock house is located some distance off to one side of the finishing house, and the cement is transported by belt conveyor through a long, upwardly inclined bridge. The cement is of course in a thin layer. Arriving at the stock house it is dumped through chutes onto another belt conveyor running the length of the stock house. This belt in turn, by means of an adjustable, sliding discharging carriage, may discharge at any desired point. The cement falls through an iron grate into the bins below. The exposure of the cement during the long travel on the belts in a thin layer, together with the shakeup experienced at the points of shift and on the grate, acts as a very efficient aerator.

An additional advantage of this method of filling the stock bins is that by moving the discharging carriage frequently from one bin to another, a very uniform distribution of the product is assured. This is actually done at the plant mentioned.

Conveniently situated with reference to the stock bins are placed the packing machines, which automatically weigh and fill either sacks or barrels, at a rapid rate. One well known machine has a capacity of four to five bags per minute. The stock bins are usually overhead so that gravity does the work.

Portland cement is packed either in barrels of 380 lbs. each or in bags of 95 lbs. each (approx.) Bags are either of heavy cotton or of paper; if of the latter they are thrown away after being used; if of the former, they are returned to the mill and used over and over again. The cotton bag is in most general use, giving best general satisfaction. An apparatus for cleaning these sacks when returned, stamping or restamping them with the name of the brand, and a cooperage outfit for heading the barrels, complete the mechanical equipment of the plant. The final bins, in which the sacked or barreled cement is stored, are handy to the shipping platform, so that loading for shipment is accomplished by hand trucking—about the only part of the entire process in an up-to-date mill that is not mechanically performed. In no other industry has continuous, automatic mechanical operation been so highly developed.

From the time the rock is fed into the crushers, it is not handled again until the finished product is delivered by the weighing and packing machines into bags or barrels. All transportation from one machine to another, or to the various intermediate storage bins, is accomplished by means of conveyors, either bucket, belt, or screw. The operation is about as continuous and automatic as it could be made.

Slag Portland Cement.—The sixth class of cement producers; namely, those using blast furnace slag and pure limestone, is of growing importance. Wherever there is a steel plant and therefore a supply of slag from the blast furnaces, these plants are sure to be established, for the slag is both a satisfactory and an economical source of raw material. The principal producer in this class is the Universal Portland Cement Co., a subsidiary of the Steel Corporation. This is not a slag cement, but a true high grade Portland cement in every respect. On the basis of cost, slag Portlands have a decided advantage over almost all other producers. The basic material, slag—a by-product of the iron industry—costs little beyond the expense of transportation. Moreover, it is already granulated so that no crushing is needed except for the small quantity of pure limestone that is used. It enters the coarse grinders directly. Another advantage of these producers is that of cheap power, the waste gases of the blast furnaces being utilized to run gas-engine-electric generators.

It should be noted that the sulphides which in a plain slag cement cause it to be debarred from important usage are in this process expelled in the operation of clinkering, going off with the other gaseous products during combustion. Thus they have no vitiating effect upon the cement.

White Portland Cement.—For many purposes a perfectly white Portland cement is desired. To produce this a careful blending of pure white limestone and pure white clay is necessary. The absence of impurities, notably iron, causes the clinkering to be more difficult. It therefore requires to be burned at a somewhat higher temperature and the resulting clinker being harder is more costly to grind. These considerations, together with the difficulty and expense of securing a sufficient supply of suitable materials, causes this cement to be much higher priced. Its present market value is about three times that of ordinary Portland cement. This of course greatly limits its use, and we find it at present employed only for artificial stone trimmings and decorative pieces, and occasionally for exterior facings. It will probably always cost more than other cement, but its cost should be greatly reduced, thereby encouraging the use of a very excellent material, and in general benefitting the industry.

Another interesting point in connection with the manufacture of white cement is that it is not possible to secure a product uniformly white with coal as fuel; for the reason that the iron in the ash of the coal causes discoloration. Consequently either gas or oil must be used.

The Wet Process.—The second and third classes, those using either **marl** or **soft limestone** (chalk) and **clay** (or shale), constitute the wet process plants. Marl, as explained previously, under composition, is in reality a soft, cheesy limestone—so because of its super-charge of water. It is a deposit of finely divided particles of pure calcium carbonate, from the skeletons of minute organisms, occurring in old lake basins. When pure it is quite white, but as usually found its color varies from light to dark gray, depending on the amount of vegetable matter present with it. This foreign matter being all organic is readily expelled in the kiln; but the water present (some 60 per cent) being mostly entangled between the microscopic particles, is not so easily removed, requiring considerable heat. This is the chief item of cost in this process. As previously pointed out, the reduced cost of preparing the raw material about offsets this, so that the cost of production for the two processes is about the same. There are a number of plants of this type in the middle west district, bordering the Great Lakes, and in the province of Ontario, Canada, but they are mostly establishments of an earlier date. For the reasons assigned previously, we do not find the present plants in these latitudes being expanded. For example, the largest marl producer in the lake region, the Sandusky Portland Cement Co., Sandusky, Ohio, is directing its expansion to the hard limestone regions of central Indiana and Illinois, which with the improvement in grinding machines have been made available. Soft limestone producers are rare, but should meet with more favor than the marl in the north; in the southern states, however, marl is about as good as any material.

The **stages in the wet process** are identical with those of the dry after the material has entered the kiln; before that essentially different, except that the final grinding of the wet material (the slurry) is accomplished in a tube mill of only slightly modified design. The slurry, as fed into the kiln being supersaturated, entails considerable extra work upon the kiln and of course delays the process correspondingly and greatly increases the cost of clinkering. Some of the water may be removed by hydraulic presses before the slurry is fed into the kiln—this considerably relieves that machine and makes for an undoubted economy.

The mixing of the materials is done in large pans, under the play of powerful mechanical agitators. The marl is introduced just as it comes from the deposit; but the clay must be first reduced to powder in a disintegrator. The marl already contains about 60 per cent water, but more is added to facilitate the mixing. The mixing apparatus resembles somewhat that used in a chocolate factory or a bakery to mix the thin cream (Fig. 13). It is simply a huge circular pan in which a pair of heavy rollers turning on a central spindle sweep around and around the enclosure, pushing and agitating, wave-like, the materials before them, speedily reducing them to a smooth cream. The material could

FIG. 13—TEN-FOOT SPECIAL MIX-
ING PAN FOR MARL AND
CLAY.

then be fed into the kilns for burning, were it not for the presence of considerable coarse grit in the marl. This requires that the sloppy mass be passed through a fine grinder, usually a tube mill. The final gauging of the mixture is accomplished while it is in the tanks, a supply of semi-liquid materials of both kinds being at hand for use as correctives.

The marl is usually conducted to the mill by pumping, compressed air supplying the power. For this purpose there are a number of specially designed apparati available. The Harris System for Marl Pumping, supplied by the Allis-Chalmers Company (see Fig. 14), embodies many unique features which are said to make it exceptionally economical. It is in use or being installed in many of the marl plants in this country and Canada. A cut is herewith given. The essential parts are the pump tanks, air compressor, automatic switch and piping. There are no floats, and no air valves outside of the engine room. After the air has done its work, it is not exhausted, but utilized in the compression cylinder, thus restoring a greater part of its energy to the compressor.

Suppose the apparatus to be in operation with the switches set as indicated in the cut. The action is as follows: The air is drawn out of the right hand tank and forced into the left hand one, thus sucking marl into the former and forcing it out of the latter.

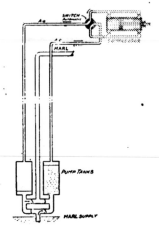

FIG. 14—HARRIS SYSTEM FOR
MARL PUMPING.

The charge of air in the system is so adjusted that when one tank is emptied, the other is just filled; at that instant the switch reverses the pipe connection and thereby the action in the tanks.

It will be noted that the air comes into direct contact with the material to be pumped, there being no plungers or pistons intervening. This makes the operation very simple and efficient. It may be adapted to most any practicable distance and lift.

The kilns used in the wet process are essentially the same as those used in the dry process; the chief modification is to increase the length proportionately and contract the diameter toward the feed-end, in order to utilize the heat generated in the lower portion of the cylinder to dry the slurry. Obviously, this contraction, together with the extra length, would have the effect of concentrating the heat otherwise wasted up the flue onto the slurry for a greater distance, thus greatly facilitating the drying. The increased cost of burning in the wet process has already been noted, and the appended table will serve to illustrate this point. It will be noted that nearly twice as much fuel is required.

TABLE SHOWING AVERAGE OUTPUT AND CONSUMPTION OF FUEL FOR DIFFERENT CLASSES OF MATERIAL AND DIFFERENT KILNS.

Process	Raw Materials	Lg. Kiln Feet	Output per day per Kiln, Bbls. Range	Average	Coal Cons. per Kiln per Bbl. lbs. Range	Average
Wet	Marl and Clay	60	60—140	85	150—250	200
Wet	Marl and Clay	80—90	80—150	100	140—220	160
Wet	Marl and Clay	110*	135	135	150	150
Dry	Limestone and Clay	60	150—200	160	90—170	130
Dry	Cement Rock and Limestone	60	180—250	200	85—160	115
Dry	Cement Rock and Limestone	80	225—300	260	85—120	110
Dry	Cement Rock and Limestone	150*	375	375	65	65

*Based on one plant—other figures based on a number of leading producers of each class. (According to E. C. Eckel.)

The above table brings out the greater efficiency of longer kilns, and also indicates the effect of wet raw materials on the cost of burning. Compiled from actual figures, it is interesting in that it shows what may be accomplished under average conditions.

The rest of the operation, as previously noted, is identical for either process.

The Power Plant.—The details of manufacture being now completed there remains to consider only the question of a satisfactory operating power for the plant. The old-fashioned way, using steam engines and a complicated system of shafting pulleys and belting, is still in vogue, but the electric motor drive is rapidly supplanting it. The substitution of the new for the old has taken place faster in this industry than in any other because the inefficiency of the old method is especially emphasized on account of the great amount of dust at nearly every stage of the process which soon clogs up the pulleys and deteriorates the belting, together with its inadequacy to the task of preserving virtually an automatic, continuous operation. For this purpose the electric drive is peculiarly well adapted, as a motor of the proper capacity and design may be direct connected at each individual stage of the process, eliminating all cumbersome shafting and belting and bringing the control of the entire plant under the direction of the operator at the switchboard. Any part may thus be thrown out without in any wise interfering with any other. It is also possible to so protect the armature of the motor, in dust proof housing, as to eliminate all interference from this source. The fact that a cement plant is in continuous operation day and night virtually 365 days in the year also makes the use of the electric drive particularly desirable.

Other points of advantage of electricity are: layout of the

plant without reference to the location of the power plant, as electric power may be conducted *ad libitum;* ability of electric motors to stand heavy initial overload, such as is required in starting the heavy grinders; ability to run continuously without special attention, reducing cost of supervision, and facility afforded in arriving at an exact knowledge of the cost of each stage of the process. Summarizing, the use of the electric drive allows the most efficient layout of the plant and gives the most efficient and economical operating power known.

The following diagram has been prepared with a view to typifying the general scheme of layout for an up-to-date Portland cement mill. Mention is particularly directed to the nice continuity of the process, which has only been made possible by the adaptation of electricity in the form of the direct drive.

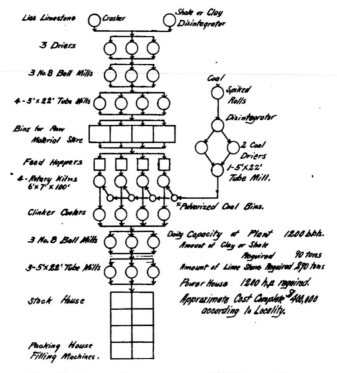

FIG. 15—TYPICAL LAY-OUT FOR A 1,200-BARREL PLANT.

To give some idea of the power required in the operation of a cement plant the following table, by courtesy of the General Electric Co., is appended:

TABLE OF HORSEPOWER, OUTPUT AND SPEED

(For some of the figures contained in the following table, we are indebted to Mr. Richard K. Meade's treatise on Portland Cement.)

Type	Size	H.P. to Drive	Revs. per Min. Pulley	Revs. per Min. Main Shaft	Output in Tons per Hour Hard Lime	Cmt. Rock	Marl Clay	Clinker	Coal	Remarks
Gyratory Crusher*	5D	25—40	400—450	200	20—30	25—35				Vertical shaft, horizontal driving pulley, output passing 1 in. screen
Gyratory Crusher*	6D	30—60	400—450	200	30—40	30—50				
Ball Mill†	7	30—40	125	21—23	3—5	4—6		2½—3		Horizontal; fed from crusher or kiln, output passing No. 16 mesh
Ball Mill†	8	40—50	125	21—23	4—7	5—8		3½—5		
Tube Mill†	5'x22'	70—80	180	21—27	3—4	4—6	8—12	2½—3	2	Horizontal; fed from ball mill, 95 per cent output pass'g No. 100 mesh
Tube Mill†	5½'x20'	80—90	180	21—27	4—6	5—8	10—15	3—4	2½	
Kominuter†	No. 66	40—55	160—175		5—7	6—8		5½—7		Similar to ball mill
Griffin Mill‡	30"	25—28	190—200	190—200	1½—2½	2—3		2—3	1½—2	Vertical; fed ½ in. crush. rock or clinker, 95 per cent output passing No. 100 mesh
Griffin Mill‡	36"	30—35	135—150	135—150	4—5	5—6				
Griffin 3 Roll‡	30"	40	150	150	4—5	5—6			4—6	
Fuller Lehigh Mill‡		30—50	210	210	3—3½	3½—4			2¾	Vertical; feed and output same as Griffin mill
Kent Mill‡		25—30	180—220	180—220	3—4	3½—4		3½—4	3—4	Horizontal; feed and output same as Griffin mill

Type	Length	H.P. to Drive	Revs. per Min.	Output in Barrels Per Day
Rotary Kiln	60 ft.	10—15	1—3	250
Rotary Kiln	80 ft.	10—15	1—3	300
Rotary Kiln	100 ft.	15—20	1—2	450
Rotary Kiln	120 ft.	15—25	1—2	580
Rotary Kiln	150 ft.	20—25	½—1	740
Rotary Kiln	170 ft.	20—30	½—1	860

*Starts light when empty; overload torque at starting if hopper contains rock.
†80 to 100 per cent overload torque necessary for starting.
‡Starts light.
§Starts with 50 to 70 per cent overload torque.

As to the total horse power required to operate a cement plant, results taken from a great many plants, indicate that from 0.8 to 1.2 H. P. per barrel per day is necessary. Thus for a 1500 barrel plant—which is about as small as it is profitable to operate —about 1500 horse power would be required. The operation when once begun being virtually continuous, the amount of power required is practically constant. This means that all units are called upon to perform up to their maximum rating continuously. There is no source of power so satisfactory, indeed so eminently fitted for such usuage, as the turbo-electric generators. Consequently, cement plants are equipping their power station with this type of generator more and more.

Laboratory Equipment.—By no means the least important part of the cement plant is the laboratory, with its equipment for making every chemical and physical test necessary on the materials at any and all stages of the process. Here competent chemists are kept employed making continual tests. In the most reliable and up-to-date plants, such a careful check is kept on the operations that the superintendent is in possession almost every hour of the exact history of the product at every stage. By this means and by the method of storing the cement so as to secure a uniform distribution throughout the bins, it becomes rather difficult, if not impossible for any inferior cement to get by and be passed on to the consumer.

Most plants now regularly test every carload shipped, doing this for the double purpose of assuring themselves of the quality and of having comparative test data with which to combat the results of field testing that may cause the cement to be rejected. All this tends to increase the reliability of the brand and its reputation with the consumer.

Relative Costs of Manufacture.—The cost of manufacture is a variable dependent on a great many factors, prominent among which may be mentioned:

(1) Nature of process, whether wet or dry.
(2) Hardness of the raw materials.
(3) Efficiency of power used.
(4) Location of plant with respect to fuel supply.
(5) Valuation of property and deposits.
(6) Efficiency of mechanical equipment.
(7) Steadiness with which plant may be operated.
Et cetera.

There are instances of Portland cement being produced for as low as 50 cents per barrel, selling costs included; and for as high as $1.25. A fair average would be between 75 and 85 cents. This is at the mill, freight rates and profit to be added for the

(For a fuller description of this question and a description of the Turbo-Generator the student is referred to the General Electric Company, Power and Mining Department, Schenectady, N. Y., who have made a special study of the application of electric power to cement manufacture and have prepared explicit literature on the subject.)

price to the consumer. The following table, prepared by Mr. E. C. Eckel, U. S. Geol. Survey, from actual statistics, presents an interesting summary of these costs:

ESTIMATE OF AVERAGE COST OF PORTLAND CEMENT MANUFACTURE.

Materials	Limestone and Clay	Cement Rock and Limestone	Marl and Clay	Marl and Clay	Marl and Clay
Process	Dry	Dry	Dry	Wet	Wet
No. of Kilns	4	8	4	4	8
Size of Kilns, feet	60	80	60	80	100
Output per day, bbls.	700	2000	600	400	1100
Cement materials	0.08	0.08	0.03	0.03	0.03
Kiln coal ⎫	0.08	0.08	0.09	0.13	0.11
Dryer coal ⎬ at $2 per ton	0.02	0.01	0.07	0.00	0.00
Power coal ⎭	0.12	0.10	0.14	0.20	0.15
Labor	0.12	0.10	0.16	0.20	0.15
Supplies, etc.	0.15	0.11	0.15	0.16	0.12
Office and Laboratory	0.05	0.03	0.05	0.05	0.04
Administration and sales	0.08	0.05	0.10	0.13	0.09
Interest, etc.	0.16	0.12	0.20	0.26	0.20
Totals, cost per barrel	0.86	0.68	0.99	1.16	0.89

CHAPTER VI.

Laboratory Work in Selection and Analysis of Portland Cement.

In the manufacture of cement products, either as concrete blocks or beams in a well-equipped yard, or with a temporary plant in the field, the *quality* of the cement entering into the combination is an exceedingly vital factor. Just as there is brought before a young man entering into concrete construction work as of paramount importance the *quantity* of cement to be used with a given amount of inert aggregate, we must emphasize the fact that a thorough understanding and an intelligent comprehension of the *quality* of the cement is fully as important, if not more so; for, while the *proportion* or *quantity* of the cement is usually determined by the specifications covering the work, and is easily secured by mechanical means, the knowledge of the *quality* of the cement requires a deeper understanding of its properties. This question is not as complex as it may seem, and if taken a step at a time as in the following pages, will soon give the student a definite grasp on the subject.

We do not mean, however, in our consideration of *quality* to make light of the question of the *quantity* of the cement used, for the questions involved and which should be considered in compiling specifications, are very important, and in general, too often overlooked. This question of the *proportion* or *quantity* of cement required for different kinds of aggregate will be discussed fully in Chapter VII, in connection with the mechancal analysis of sands and aggregates.

Apparatus required.—In order to properly test the materials to be used in concrete, and also to test concrete after its manufacture, a certain amount of apparatus is required. At the very outset, two sieves are necessary for determining the fineness of the cement. This in itself is a very important point, for upon the fineness of the cement depends its strength or bonding power, and cement which does not pass the sieves as required, will not give results to be obtained from cement of the required fineness. Consequently a coarsely ground cement is an inferior article and has not the same efficiency as an equal amount of properly ground cement. Hence the necessity for the screen test.

[NOTE: *This chapter is a revision of the text used by the students of the Institute of Applied Concrete Engineering in field and home laboratory analysis of cement. The detailed description of the special scale is omitted from this volume When "Institute" is mentioned, it refers to the above mentioned Institute of Applied Concrete Engineering.*]

To carry out the experimental work as detailed in the following sections, the following equipment will be necessary.

1 balance, capacity 4 ounces, beam graduated to .01 oz. and .01 gram.

4 scale weights, 1 oz., 2 oz., 25 gram and 50 gram.

14 grading sieves, 4", brass, with tin sides, No. 2 to No. 200.

1 testing machine, capacity 4,000 lbs. Tensile, transverse and crushing.

1 block pressure mold, $\frac{1}{2}$" x 1" x 2".

1 standard single briquette mold, 1" x 1" x 3".

1 density gage, $\frac{7}{8}$" x 8", brass.

12 test tubes, 6" x $\frac{3}{4}$", plain.

2 test tubes, 6" x $\frac{3}{4}$", graduated to c. c.

1 test tube rack.

1 pipette, 8" x $\frac{1}{4}$".

1 graduated measuring glass, 200 c. c.

1 wash bottle, 1,000 c. c. complete with tubes.

3 feet glass tubing, $\frac{1}{4}$".

4 feet rubber tubing to match.

2 feet $\frac{1}{4}$" glass rod.

1 mold for 5—1" test cubes.

1 mold for 2—2" test cubes.

1 volume gage.

The Use of The Apparatus.

The Density Gage.—Having weighed out one ounce of the cement which is to be tested, it becomes necessary to ascertain its volume in cubic inches, together with its weight in pounds to the cubic foot. The gage shown by Fig. 16 consists of a cylindrical body G, attached to a broad base which prevents it from being easily overturned. A plunger H, which fits snugly the interior of the instrument, is graduated with three scales as shown at a, b, and c. Scale a gives the volume of the material in cubic centimeters. Scale b gives the volume in cubic inches, and scale c gives the weight to the cubic foot. A removable washer, d, made of hard rubber, or a very dense wood, is attached to the lower end of the plunger, and as wear occurs through use, this washer should be replaced whenever the material to be measured finds its way past the washer. Necessarily there will be wear in any instrument, and the simple replacing of this washer when worn restores the gage to its original exactness. By placing one or more thicknesses of paper or cardboard between the washer and the end of the plunger, the upper end of the several graduations may always be brought even with the top of barrel G.

Using the Density Gage.—To use the density gage for finding the volume of any sample of sand, cement or similar material is quite simple. First, test the gage by inserting the plunger to see if the upper line of the graduations coincide with the top edge of the barrel. If the plunger does not go into the barrel far enough, there is something inside which should be removed before making the test. If the plunger goes in too far, the washer *d* needs renewing before accurate measurement can be made.

Place a weighed quantity of material in the barrel of

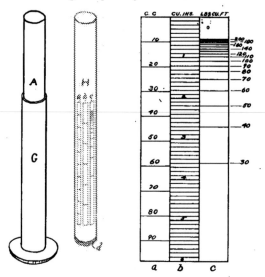

FIG. 16—DENSITY GAGE.

the gage and give it a whirl between the palms of the hands to level off the top of the material inside the gage. Next, drop the plunger down upon the material and note the reading upon the graduation *a*, *b*, or *c*, as may be required by the work in hand. Assume that one ounce of cement has been weighed and placed inside gage barrel G. The plunger is dropped into the barrel and projects far enough, after it reaches the cement, to expose the graduation upon all the scales, above the 1.45 cubic inch mark.

Thus, one ounce of this particular sample of any Portland cement occupies 1.45 cu. ins. loose, as poured into the gage. By glancing across to the other scales, it is seen that the cement has a volume of about 24 c. c. (cubic centimeters). By glancing to scale *c*, it will be noted that the weight to the cubic foot is about 75 pounds.

There are two ways of taking the weight of granular substances: When loose, and when shaken down. In all the

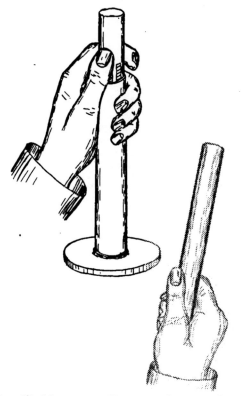

FIG. 17—MANNER OF USING THE DENSITY GAGE.

tests and calculations found in these lessons, the weight is taken with the sample shaken down, so that no decrease in volume is observed after two minutes shaking. This may be called "shaken to refusal," as the volume refuses to decrease by further shaking.

Shaking the material in the gage may be accomplished by grasping the barrel with the three last fingers and the palm of the hand, with the thumb and first finger holding the plunger firmly down against the material in the gage. A bit of broom stick about 18″ long is brought into play, as shown by Fig. 17. which also shows plainly the manner of holding the gage barrel and plunger. After each three of four blows, the plunger should be raised and given a slight turn, one way or the other, in order to shake down beneath the plunger, whatever material may work up between it and the barrel when the gage is struck by the stick. In gaging very fine material, there is a little trouble in this direction when the operation is commenced, but after the sample settles down a little, there is no further trouble. But the constant lifting and turning of the plunger makes sure that no material is caught between plunger and gage-barrel.

When the gage is emptied after a measurement has been made, always push in the plunger and see if it goes down until the graduations are entirely out of sight. Some of the material is liable to adhere to the plug in the bottom of the gage and if not discovered and removed before another measurement is made, it may seriously affect the accuracy of both the test from which the material came, also the next test which may be made. There must be ceaseless vigilance in performing each and every weighing or measuring operation, that no material be lost or that no foreign substance gets into the sample being acted upon. In all the operations we shall go through with, very small quantities of material are to be used, in order to save labor in sifting, weighing, testing and otherwise manipulating. When tests are made in ounces, and multiplied by 32,000 to handle material in tons, it will readily be seen that considerable accuracy is necessary in order to obtain results which can be duplicated in the factory from the tests made in the laboratory. There is little trouble in securing the same results in either place, if reasonable care is taken that no material be lost or displaced during the testing, and that the weighing and measuring be fairly accurate. Be exact, therefore, and save much trouble and disappointment later on.

When cement has to be measured in the density gage, or when very fine sand, say through 200 mesh, has to be treated, considerable care is necessary in placing the plunger in the gage barrel. If the plunger be forced down suddenly, a considerable quantity of very fine material will be forced out with the rush of air beside the plunger, and the quantity of material thus forced out will be lost and lessen the accuracy of the test. Insert the plunger slowly and gently when fine material is being handled. The barrel is full of air, which must all escape around the plunger as it slips into the barrel of the gage, therefore give the air time enough to pass out so slowly that it will not take a lot of fine particles with it.

Density of Portland Cement.—Upon shaking down the ounce of Portland cement as described, it will be found to measure about 0.9 cubic inch, or 14.5 c. c., and to weigh 117 pounds to the cubic foot. It is stated "about" 0.9 cubic inch, about 117 pounds, etc., for the reason that the sample tested by the student may vary in volume a trifle, from the samples tested when the figures above given were obtained. These figures are accurate for the samples tested, within the limit of error of the instruments, which is probably a very small fraction of 1 per cent. Thus, while the measures and weights cited in these pages are accurate for the samples from which they were taken, they can only be compared with the student's weighing and measuring by saying "about."

Fineness of Portland Cement.—In order to pass the test for being pulverized to a sufficient degree of fineness, cement shall, according to the requirements of the American Society for Testing Materials, pass at least 92% through a 100 sieve, and 75% through a 200 sieve. Cement should be dried before it is tested for fineness. Weigh one ounce and place it in an agate or granite-lined dish and place where it will be kept at 212 degrees (Fahrenheit thermometer is used in all tests described in these lessons) for at least two hours to drive off any moisture which it may contain. And cement nearly always contains some moisture and the sample under test will probably lose at least 1% in weight during the drying process. Record the loss by drying, for in later tests to be made, the amount of moisture present in the cement plays an important part.

In a regular test for fineness, the weighed quantity is placed in the 200 sieve and shaken until less than 1/10 of 1% passes through the sieve in one minute. In recommendations by a committee of the American Society of Civil Engineers, the following directions are given for sifting cement:

The thoroughly dried and coarsely screened sample is weighed and placed on the No. 200 sieve, which, with pan and cover attached, is held in one hand in a slightly inclined position, and moved forward and backward, at the same time striking the side gently with the palm of the other hand at the rate of 200 strikes a minute. The operation is continued until not more than one-tenth of 1% passes through after one minute of continuous sieving. The residue is weighed and placed on the No. 100 sieve and the operation repeated. The work may be expedited by placing in the sieve a small quantity of large shot. The result should be reported to the nearest tenth of 1%.

Cement which fails to pass 75% through the 200 sieve should not be used. Upon the fineness of the cement depends its power of uniting together the particles of aggregate, and that portion of the cement which fails to pass the 200 sieve has only about the value of so much sand. Coarse cement adds practically nothing to the uniting or holding power of that substance.

For the purpose of testing in general, the sample of cement may be placed in the coarser of a nest of sieves and separated into several grades, ranging from, perhaps, "on 50" to "through 200." This means that some of the cement particles are so coarse that they fail to pass through a sieve having 50 openings to the linear inch (2500 to the square inch), while the bulk of the cement passes through the 200 sieve which possesses 40,000 openings to the square inch. The sieves as nested for cement testing shall include Nos. 40, 50, 60, 80, 100, 120, 150 and 200—eight sieves in all—a description of which will be found in the following chapter (VII).

The cement used in the tests to be described in the following lessons, was separated by the above named sieves as follows:

Through No. 40   a trace of fine silica sand.
"   "   50   0.6% = 6/10 of 1%.
"   "   60   0.9 "
"   "   80   4.4 " = 4 4/10%.
"   " 100   6.4 "
"   " 120   4.9 "
"   " 150 11.4 "
"   " 200 71.4 "
                 ———
Total,   100.0%

The result of the sieving shows that 5 9/10% was too coarse to pass the 100 sieve, 22 7/10% was caught between the 100 and 200 sieves, while 71.4% passed the 200 sieve. This is a fairly good sample of cement, though the fineness is 3 6/10% shy. But the amount through 100 sieve is 2 1/10% greater than is called for.

Strictly according to specification requirements, this cement could be condemned because 75% did not pass the 200 sieve, only 71 4/10% going through. Good concrete, withstanding a pressure of 2450 pounds to the square inch when 28 days old, was made from it, less than 12% being used and the absorption was less than 7% of the weight of the block. It is that portion of the cement which passes through the 200 sieve which is of value to the cement user. The remainder of the cement acts as a mere filler, taking the place of so much sand, as will hereinafter be shown. The several grades of cement may be put into a test tube with thin cork sections between, to form an interesting object lesson in the value of cement. A test tube thus filled will be illustrated and described later on pages 90 and 91.

It has been shown that a certain sample of cement weighed 117 pounds to the cubic foot. As 1 cu. ft. of water weighs 62½ pounds, the specific gravity of the cement may be taken as $117 \div 62.5 = 1.87$. But as cement is supposed to possess a specific gravity of about 3.1, there is evidently something wrong in the estimate. The difference may be laid to the amount of void space between the particles of cement, which in this instance is found to be 37%, leaving 63% of solid material in the volume of cement. To make corrections for the voids, there will be $1.87 \times 100 \div 63 = 2.97$, which is a close approximation of the specific gravity of a Portland cement.

To ascertain the Specific Gravity of any sample of cement, there is necessary, in addition to the weighing scale, only a test tube and a bit of glass tube in the form of a "pipette." A test tube is shown—several test tubes, in fact—by Fig. 18, which also illustrates a method of ascertaining the specific gravity and the percentage of voids in any substance which can be placed inside a three-quarter-inch test tube. Other test tubes, larger or smaller, may be obtained, but a tube more than ¾″ in diameter cannot be readily covered with the thumb when it is necessary to shake the contents of the tube. A less diameter than ¾″ is not always convenient when gravel or broken stone, ½″ in diame-

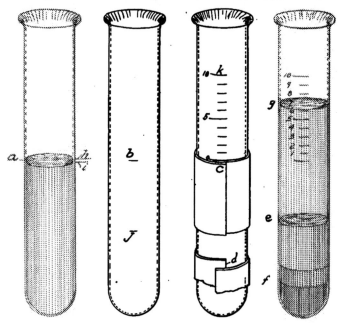

Fig. 18—Test Tubes Arranged for Specific Gravity Determination.

ter, is to be tested for voids, hence the ¾" size is adopted as the most convenient tube for use with these lessons.

Some test tubes come graduated to cubic centimeters, but they cost more and are not always to be obtained. All the graduations necessary may be easily placed upon a test tube by means of a fine file. A three-cornered file may be used, but a smooth-cut "knife" file is better. A half round file, dead-smooth cut, is also an excellent thing for marking test tubes, and other laboratory glassware.

Select a tube and weigh it. Clips are attached to the suspension wire of the scale scoop for the purpose of holding upright a test tube when placed in the scoop. Balance the tube accurately, and it is well to scratch on the tube its weight in grams. It often comes handy to know the weight of a tube and if the weight be scratched upon its surface, the figures will be at hand when wanted. After the tube has been balanced in the scoop, slide the beam weight out 20 grams more, and fill the test tube with water until it just balances the scale again. As stated before, water at 60 degrees may be used with only a very slight error, and the 20 grams of water placed in the test tube will occupy just 20 cubic centimeters of space. Thus the tube filled with water to a contains 20 grams of that fluid, and the volume of the tube up to mark a is 20 c. c.

The concave water line observed when a tube is partially filled is due to capillary attraction. Water always rises along the wall of a tube, higher than it does in the middle, and the smaller the tube, the higher will be the water along the side of the tube. In filling a tube to a given point, there are two ways of so doing. Either to fill the tube until the liquid on the side of the tube comes level with the mark on the tube, or else until the liquid in the center of the tube comes even with the mark. The first method is used in all cases in these lessons. When the tube is filled until the water in the center is level with the mark, the difference is equal to the space h, i.

It is well, when putting the 20 grams of water into the test tube, to put in 10 c. c. at a time, and mark both levels as shown at b and j. The tube can then be used as a measure for either 10 c. c., or 20 c. c., which will prove very convenient for one of the operations described in succeeding paragraphs.

The points b and j should be marked plainly with a file, as already noted, and the marks should be at least ¼" in length. It is not well to mark entirely around the tube, as breakage is apt to occur where marks extend entirely around. It is, however, necessary that all marks be carefully made and square with the length of the tube. When a graduation leads up or down, instead of squarely around the tube, then the graduation is worthless, and should be discarded and a new attempt made upon a fresh tube.

A very easy way of squaring a mark with any tube is shown at *c*, where a bit of thin brass, tin or thick cardboard, is wrapped around the tube and brought even with the zero mark 0, as shown. Use the wrapping as a guide for the file and the mark must come square with the tube. Be sure that the wrapping is even at the ends. When one edge is up and the other edge down, as at *d*, it is impossible to make a mark perpendicular to the tube and the file will be cutting a thread instead of marking squarely around the tube.

A tube or two should be graduated as shown at *c, k*, a mark being made at each c. c., or 10 c. c. All the tubes may thus be graduated, if desired, and they will come very handy at times, still, if lack of time prevents, one or two tubes thus graduated will answer. In finding the marks between 0, and 10, on *c, k*, two courses are open. By one method, 20 c. c. of water may be weighed into the tube and mark 0 made as described; then another centimeter of water may be added by weighing in one gram more of that liquid, and the second mark made in the graduation. Continue weighing in water, one gram at a time, and

FIG. 19—A PIPETTE AND METHOD OF MAKING.

mark each new level until 10 c. c. has been added, which will bring the graduations to 10 at *k*. By the other method, weigh in 20 grams of water and mark 0, at *c*, then weigh in 10 grams more and mark 10 at *k*. Mark on a piece of paper the exact distance between 0 c. c. and 10 c. c.; space into 10 equal parts, paste the paper beside *c, k*, then bring *c* to each line in turn and transfer same to the glass with the file, after which the slip of paper is to be washed off and the graduations will be found complete as shown.

For filling a test tube with water to any line or mark, some arrangement is necessary whereby a very minute quantity of water may be added to, or taken from, the test tube. An attempt to fill a tube, or any other vessel with water to a certain mark, either by pouring water in or out, can only lead to a waste of time and inaccurate work. The pipette should always be used for this purpose. This instrument is shown at *a*, Fig. 19.

As they break easily, the student should arrange to make them as required, the method of so doing being very simple and fully shown by Fig. 19, in which a completed instrument is shown at *a*, and a bit of glass tube from which one is to be made is represented at *c*. Heat the tube in any very hot flame. An alcohol lamp, a blacksmith's forge, or even an ordinary coal fire, will do the trick. The lower part of an ordinary Welsbach gas fixture is in reality a Bunsen burner, such as is used in some of the best equipped laboratories. The student can very easily secure and equip this by simply removing the top or mantle part of the Welsbach. This will work best when placed on a low table. When hot, pull lengthwise of the tube and stretch it as shown at *d*, until it is only about 1/16″ in diameter at that point. When cold, break it at *d*, and the instrument is complete.

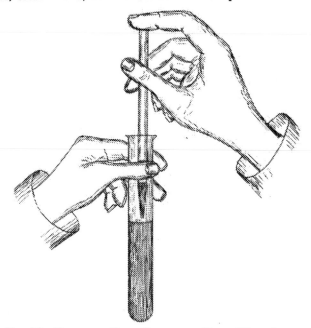

Fig. 20—Filling a Test Tube to a Mark With Liquid.

The proper method of filling a tube to a given mark or weight is illustrated by Fig. 20. An approximate quantity of liquid is first poured in, then the weight is taken so far as to ascertain if there be too much or too little water in the tube. To remove some of the liquid, lower the pipette until the small opening is some distance below the surface of the water in the tube. When the estimated quantity of water to be removed has passed into the pipette, place the finger over the upper end of the pipette and remove that instrument, together with the water contained in it. The opening in the pipette being very small, the capillary attraction is great enough to prevent any of the water running out or any air getting in, therefore the contents of the pipette will stay inside until the finger is removed from the large end of that instrument. It is now possible to add to, or remove from the test tube, a minute quantity of water, thereby quickly bringing the surface to a mark, or the scale to a balance.

To secure accuracy in specific gravity and void determination, great care is necessary to fill the tube exactly to the mark, and to remove the liquid exactly to the same point when making the determination. The distance between the 20 c. c. mark and the 30 c. c. mark is only a trifle over 1½" and as this distance represents 100%, it is evident that an error of 1/100" in filling or emptying the tube to the mark, will affect the accuracy of the result six-tenths of 1 per cent. It is, therefore, necessary to work with all the accuracy possible. With care, the specific gravity may be determined within one-fourth of 1 per cent as a general result.

The specific gravity of a sample may be obtained in two ways. As shown by Fig. 18, the percentage of solid may be read directly from the scale on the test tube, the distance the water rises above the 20 c. c. mark being the number of cubic centimeters of solid in the sample of cement. Consequently the distance between the water level and the 30 c. c. mark at **k** (also marked 10) represents the volume in c. c., of the voids. By dividing these readings by 10, (10 centimeters of material is being tested, and if some other quantity has been used, divide by that number), the percentages are obtained direct. Thus, when the 10 c. c. of cement has displaced 6.3 c. c., the percentage of solids is 63% and the voids equal 100 — 63 = 37%.

The second method, which is used when the test tube has not been graduated between 20 and 30 c. c., is to proceed as before until the time for making the reading is reached. Then, as there is no scale to read, remove from the test tube, all the water above the 20 c. c. mark and place the water removed in the scoop of the weighing scale. The pipette will do the removing act readily, and to remove the water more quickly, place the lips over the end of the pipette, suck the instrument nearly full of water, place the tongue over the end of the tube, quickly replace the tongue with a finger, keeping the small end of the pipette in the test tube all the time, then when the finger is safely and fairly over the large end of the pipette, carry that instrument to the scale and let its contents run out into the scoop. After the water is nearly down to the 20 c. c. mark on the test tube, proceed as in Fig. 20 until the water level is on the mark again, when the water in the scoop may be weighed and its weight in grams treated in the same manner as was the amount read from the scale, Fig. 18, in cubic centimeters.

It is one of the beauties of the metric system that the

weight and volume of water interchange so readily, thereby making it very easy to work in the metric system of weights and measures. It may be a surprise to the student to learn that cement contains 37% of voids. It is even so, and in repeating the test, a volume of cement taken from some other part of the same package may even show a percentage of voids amounting to 40%, 44% or even 46%. There is a great difference in voids in cement, also in sand, gravel or broken stone. To obtain a true indication of the actual voids in a lot of material, it is necessary to make determinations from many parts of the material, add the several readings and divide the sum by the number of determinations made. The result will be the average percentage of the voids, specific gravity, or whatever the tests are made to determine.

When results do not agree, and they seldom do, closer than 2%, the student should by no means be discouraged. He may be right, and no error can be found in the work. Just repeat the determination, and take an average as above described. While single tests by different persons may fail to agree, and while different tests by the same person also show a considerable variation, it will be found that the average result obtained by one man will agree very closely with the average obtained by other workers, provided that the operations were correctly performed and uniform accuracy observed in weighing and measuring.

The percentage of voids has been ascertained, but as yet the specific gravity of the cement sample has not been found. To obtain the percentage of solid matter in a given quantity of cement, it has been found necessary to divide its displaced volume by its apparent volumes as determined in the density gage. In the case of the cement sample, the displaced volume, 6.3 c. c., was divided by the apparent volume, 10 c. c., and the result was 0.63, the percentage of solid matter in the 10 c. c. of volume occupied by the cement when shaken down to refusal. As shown previously, where an approximation was made of the specific gravity of this sample of cement, if the weight be divided by the actual or displaced volume of any substance, the quotient will be the specific gravity of that substance.

The specific gravity of any substance, as the term is used, means that for the same volume the weight of the substance is so many times the weight of water. We have found that the weight of 10 c. c. of water is exactly 10 grams, hence' the specific gravity of water is $10 \div 10 = 1$. We find that the weight of the 10 c. c. of cement is 19.53 grams, and dividing the weight by the volume (6.3 c. c.) the specific gravity is obtained. Thus: $19.53 \div 6.3 = 3.1$, which is the specific gravity of Portland cement called for by most specifications. This means that the cement is three and one-tenth times as heavy as water.

In placing the weighed and measured sample of cement inside the test tube for void and gravity determination, care must be taken that none of the cement is spilled, and that no air bubbles are caught and carried down under the water. It is best to shake the cement slowly into the test tube, either from the weigh-scoop or from another test tube, and a funnel should always be used to conduct the material from scoop to test tube or density gage. The funnel A, illustrated by Fig. 21, is made from a single piece of thin brass, the body *a* being rolled up into

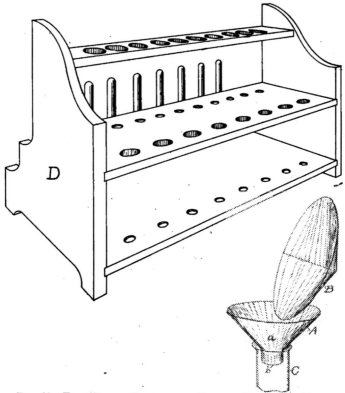

FIG. 21—TUBE RACK—FUNNEL FOR FILLING TUBES AND GAGE.

a cone and soldered or riveted. Then the short nozzle *b* is formed by hammering upon a round rod until a half inch of the cone *a* has been drawn out parallel with the center line. The illustration shows the manner of using, the funnel A being held over and in the tube by one hand, while the scoop is tilted by the other hand. In this matter the scoop can be rattled against the funnel to shake out any adhering particles of material from either scoop or funnel. The rack shown at D should be used for holding test tubes, which should be placed in the holes while in use, and hung on the pins when not in use, especially when wet.

The minute directions given for seemingly very simple operations are more important than they appear at first sight. To make accurate tests and determinations, great attention must be given to the smallest detail. The large things will take care of themselves, and accurate testing is nothing but the most scrupulous attention to minute details. Therefore, study the methods given, and pay strict attention toward doing things as described until accuracy and attention to minute detail becomes a matter of course. Until that point is reached, the student will obtain anything but accurate or uniform results. But, as soon as attention to detail becomes a habit, there will be no more trouble from provoking little happenings whereby a drop of some liquid is lost, or a lump of sand or cement adhering overlooked in a corner, impairing the result of a test. Take good care of the detail and the road is an easy one. Neglect detail, and results are just like concrete blocks made in the same manner, i. e., not worth carrying home.

With the cement sample safely in a test tube, together with 20 c. c. of water, see that no air bubbles have been carried down with the cement. It is best to place the thumb over the top of the test tube, which is then inverted and shaken until all the lumps have been washed to pieces and the air-bubbles all dislodged. To prevent some water being lost upon the thumb when placed upon the tube, the thumb should be wetted before it is placed over the tube. To prevent any of the cement from being deposited upon the thumb, hold the tube still a few seconds after shaking it and as soon as the cement settles a little in the tube, another slight shake or two will wash down any particles which may adhere to the thumb, which may then be removed and the tube placed in the rack to allow the contents to settle, after which the water above the 20 c. c. mark may be measured or weighed as described in following pages.

The method of void determination described above is well adapted to use with sand, gravel or similar substances which are not soluble in water. The term "soluble in water" is used to designate solids which are disintegrated and held in intimate solution. Thus, sugar and salt will dissolve in water until a certain quantity has been taken up by the liquid. Clay seems to dissolve in water, but it does not. It simply is mixed up with the water and will settle to the bottom upon being allowed to remain at rest long enough. Some kinds of clay are very hard to settle. They are about the same weight as water and stay floating around in it a very long time before they go to the bottom. Such substances are said to be "suspended" in the water, or to be "held in mechanical suspension." The sugar and the salt are said to "enter into solution" and the union with water is a chemical one, while the union with clay was a mechanical one.

It is very evident that we could not determine the voids in sugar or salt by the method described, and therefore the method, while it is accurate in itself, must be modified for use with substances which will dissolve in water. Cement may, and does dissolve partially in water. A portion of the lime contained in cement enters into solution, while the remainder of the substance remains in mechanical suspension and its voids are measured.

It is important to know what portion of the lime has been lost in the water, and it is also important to be able to make a determination in some manner which will not be open to an error of unknown dimensions. To secure accuracy in the determination, it is evident that some liquid must be used which will not dissolve any of the lime or other substances of which cement is composed. Such a liquid is common kerosene oil. Lime is not soluble in kerosene, and therefore kerosene may be used for determining the voids of lime, cement, and all similar substances which will not dissolve in it.

It will be well to make two more determinations of voids and specific gravity, one with water, the other with kerosene oil, in order to compare results and to see how great is the error when the determination is made with water. In order that the conditions may be as nearly alike as possible, use portions of exactly similar cement for both determinations. For thus purpose measure out 20 c. c. of cement and mix it thoroughly to make sure that it is as uniform as possible. Having found that the cement at hand weighs 19.53 grams to 10 c. c., double that amount may be weighed out at one time, mixed thoroughly and divided into two equal portions of 19.53 grams each. These portions may be shaken in the density gage to make sure that the cement is uniform with that previously tested. If it will not shake down to two portions of 10 c. c. each, make the necessary changes so it will shake down, then weigh again and take the new weight instead of the 19.53, used heretofore.

It will be assumed that the weight remains constant, as was the case with tests actually made for use with this paragraph. The record of the test, taken from the record book (every student should establish a record book as described later on, (page 104), and record therein every measurement, every weighing, and every operation, its conditions and its results. Neglect to do this will many times cause loss of valuable data and the waste of much time and labor) shows that the weight remained constant at 19.53 grams. The water test showed a displacement of 6.05 grams or cu. cms. of water, thus fixing the solids at 60½%, and the voids at 39½%. Thus the voids are 2½% greater in this portion of cement. This may be expected, as the voids frequently vary as much as 5% in the same barrel of cement.

The determination with oil gave the following results: "Oil displaced, 4.95 grams." This would lead the student to assume that the sample tested had a solid volume of 49½% and voids of 50½%. But there is another factor to be considered.

Oil is lighter than water. It does not weigh as much to the cubic foot or centimeter, hence more volume has been displaced in the test than is accounted for in the 4.95 grams displaced by the 19.53 grams of cement.

. A test should be made to determine the specific gravity of the oil used. Select one of the marked test tubes and fill it with kerosene oil to the 20 c. c. mark. Here is a tube which has its weight, 20.75 grams, scratched in the glass by means of a little diamond in a scarf pin. Filled with oil to the 20 c. c. mark, the tube weighed 36.95 grams. Deducting the weight of the tube, there remains 16.2 grams as the weight of 20 c. c. of oil, or 8.1 grams for 10 c. c. Dividing the volume by the weight—to obtain the specific gravity—it is found that the specific gravity of this particular oil is $8.1 \div 10 = 0.81$. The specific gravity of kerosene varies between 0.80 and 0.83, and every time a new lot of oil is used, its specific gravity should be determined as above. It is well, to save labor, to secure a couple of quarts of as heavy oil as possible—there is less smell from a heavier oil, hence less danger of error through evaporation as would be the case were gasoline used—and store the oil in a Mason fruit jar, where it will be easily accessible and free from contamination.

The specific gravity of the displaced oil having been found to be 0.81, the volume may be readily calculated. As a cubic centimeter of oil weighs only 0.81 gram, instead of 1.0 gram as is the case with water, then there must be in the displaced oil, $4.95 \div 0.81 = 6.11$ c. c., or the solid portion is 61.1%, leaving the voids in the cement equal to 38.9%, against 39.5% as determined by the water method, a difference of 6/10 of 1 per cent, showing that the amount in question of soluble matter was dissolved and taken up by the water during the water test.

The importance of this discovery will not be apparent to the student at this time, but it will be appreciated later. And as the kerosene determination is much more unpleasant to make than the water determination, it will be frequently found convenient to test the cement by the water method and then diminish the result obtained about 1% to approximate the oil determination. It will be well for the student to make the following experiment in order to ascertain what the 1% consists of which is absorbed by the water during that determination, also to put the cement composition in such shape that it may readily be seen, as noted in paragraph 108.

The mechanical composition of cement may be made very apparent by the following arrangement of about one ounce of that substance. After drying a quantity of cement with a dry heat, of at least 212 degrees, weigh out 1 ounce. Then measure out about 16 ounces of as pure water as possible. Rain water is preferable and water condensed from steam is best of all. Stir the cement into the water until no lump is left. A very good way is to sift in the cement through a No. 50 mesh, stirring the water while the cement is being sifted in.

After stirring thoroughly for at least one minute, set one side to settle, covering closely to keep the air from it as much as possible. After standing at least one hour, pour off the clear liquid into a clean agate or granite-ware dish, allow it to settle again, and pour into a third dish. Any residue found in the second dish is to be returned to the first dish, and if the least sediment appear in the liquid, the clear portion shall be poured off very carefully into a clean dish, again and again, if necessary, until nothing but very clear liquid is in the last dish. Whenever a dish is used, it must be rinsed with clean water, and every particle of sediment returned to the first dish, in order that nothing may be lost, an occurrence which would destroy the accuracy and the value of the entire experiment.

After evaporation to dryness, weigh the residue from the poured-off liquor. It is a white, or grayish white powder which adheres to the bottom of the dish and a little care is necessary to scrape the dried-on sediment off the dish without losing any of the material. The long thin pallet knife supplied with the outfit, is the proper instrument for removing the sediment from the dish. After scraping it carefully into the weigh-scoop, its weight will be found to be about .015 ounce, about one and one-half spaces on the ounce scale.

Next, weigh the residue from which the liquor was poured off, it having been dried out at the same time the water was evaporated. The residue will be found to weigh about .995 ounce, or, one-half of one per cent of the cement has been dissolved out by the water in which it was stirred up. But it was found that the residue from the poured-off water dried out to .015 ounce, or .01 ounce more, than there is missing from the cement. This apparent discrepancy is due to the fact that the lime and other soluble matter which was dissolved out of the cement, has absorbed or taken up water or carbon, or perhaps some of both. The lime in hydrating, would take up at least 25% of its weight of water and some carbon dioxide (sometimes called "carbonic acid gas") will be taken up if the drying out was where flame or hot air from the fire could get at the inside of the dish in which the evaporating was done.

There is another source from which some of the excess weight might come and that is from the water in which the cement was stirred. Place an amount of water, equal to that used for soaking the cement and washing the dishes, in a clean dish, evaporate and weigh the residue if any and deduct the weight from the .015 ounce. The residue found in this manner comes from lime or other substances contained in the water, and it illustrates, pretty forcibly, the necessity for using distilled water for all tests and experiments, made on a small scale.

It has been found that .995 ounce remained of the 1.000 ounce of dried cement originally taken for treatment. 1% of water, at least, is usually contained in cement as received in the bag or barrel, and when the sample under observation was weighed previous to the soaking test, it was found that the loss in drying was 1% in this instance also. Thus, of the commercial cement such as would be used for concrete, 1.01 ounce must be taken to make up the 1.0 ounce after drying.

After the washing-out process described on page 89, it was found that .995 ounce remained, the actual filling volume of the 1.1 ounce of cement is, then: $.995 \div 1.01 = 98.5\%$ of its original weight. This fact must be remembered when proportioning cement and aggregate so that with 12% of cement, only 11.83% goes to increase the volume of the block, the remaining 1.5% being chemically combined and not appearing in increase of volume. In addition to this, bear in mind that the cement is about 40% voids, thus bringing the actual solidity of the cement used in a block down to $.1183\% \times .60 = .07098\%$, or nearly 7.1%.

The reader should carefully study the fact that 12% of cement adds only 7.1% to the solidity of any block, and when the claim is made by some earnest but uninformed worker that he "adds cement enough to fill all the voids," the student may attach to the statement the proverbial "grain of salt," and lose no time in arguing the question that voids can be filled with a substance which is of itself 40% air-space.

A cement record may now be made of the preceding test which will prove both instructive and interesting. Some "parallel" corks should be procured, which will just fit inside the test tubes. Taper corks will answer, but they are not as good as the "parallel." With a very sharp knife, slice off three slices of cork 1/16″ thick, as shown at a, b, and c, Fig. 22, sketch A. If the cork slices do not quite fill the tube, lay each slice flat on a piece of smooth iron and hammer lightly. The slices will widen considerably under this treatment.

Place in the bottom of a test tube, as at e, sketch B, the lime residue obtained by evaporating the cement wash-water. Push down on top of the residue, cork a, then sift the .995 ounce of residue through the 100 and 200 mesh sieves. That portion which will not go through the 100 sieve, is placed in the tube (Fig. 22) at f, then cork b is pushed down; the material from sieve No. 200 is placed at g, and cork c shoved into place. Next, the remainder of the cement, that portion which passed through the 200 sieve, is placed as at h, and the longer cork d, is forced down upon it.

We now have the record completed. It is well to pour a little shellac varnish over cork d, to prevent atmospheric moisture from swelling the cork and cracking the tube—something which frequently happens if the cork be forced in tightly and not protected from moisture in the air. A little shellac, paraffine, or even oil, will furnish sufficient protection against moisture.

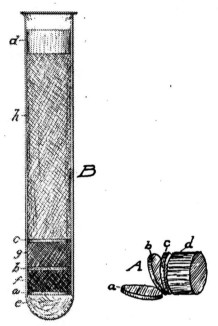

FIG. 22—A CEMENT RECORD.

The upper and light colored portion of material in the
tube (h) represents 71% of the cement and is all that does any
good in fastening the grains of gravel and sand together. The
white portion in the bottom of the tube is some of the material
which actually does the cementing by combining with the silica
of the aggregate, while the coarse portions in the middle of the
tube, f and g, are practically worthless as far as any cementing
is concerned, and should only be considered, in concrete making, as so much sand. Experiments made with this portion of
the cement removed, showed very little benefit from the presence of coarsely ground portions of the cement, thus proving that
the more finely ground the cement, the better it is as far as
its cementing properties are concerned. .It also goes to prove
that an excess of finely-ground cement in a mixture, can only
decrease the weight per cubic foot, decrease the strength, and
increase the percentage of absorption. Hence: the value of a
concrete is not only enhanced by using a proper proportion of a
properly ground cement, but an actual saving in amount of cement required is possible. It should also be apparent that an
excess of coarsely ground cement is less harmful than an excess
of finely ground cement. What is the proper proportion of any

cement? Whether finely or coarsely ground—that is one of the problems for the student to solve? He will also ascertain the proper amount of sand and of gravel which must be used with each grade of cement in order to obtain a mixture of maximum density.

But little attention need be paid to the testing of the chemical properties of cement, though they will be briefly de-

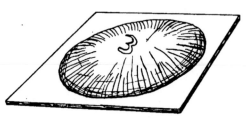

FIG. 23—CEMENT PAT FOR TESTING SOUNDNESS AND CONSTANCY VOLUME.

scribed in the following paragraphs. The fineness of the cement is of the greatest importance to the cement user.

The activity, or time of setting is not important in certain lines so much as in others. The soundness or constancy of volume of the cement must always be taken into consideration. To test this property of the cement, procure a number of pieces of window glass at least 4″ x 4″ and clean them with Soapine, Pearline or their equivalent. There must be no grease or finger-marks on the pieces of glass to prevent the cement from adhering.

Mix up at least 9 parts of 50 grams of cement each and spread the cement upon the glass about ½″ thick in the middle and tapering to a very thin edge around the entire circumference of each pat, as shown by Fig. 23. Work the cement well down at the edges with the pallet knife and do the work quickly, before the cement begins to set.

It is very important that the parts be all mixed with an equal and correct proportion of water. Each variety of cement may require a different percentage of water, and in making pats, the cement should be mixed to that degree of wetness which is known to cement testers as: "Normal Consistency." In laboratories, an instrument known as the "Vicat Needle" is used to determine both the initial set and the time of setting, otherwise known as the time from the addition of water until the time when the paste begins to harden and until it becomes fully set.

An approximation of the time of initial set may be

obtained from one or two of the pats noted on page 92, by means of a very crude and rudimentary substitute for the Vicat needle. Cut off a bit of ½″ round iron rod and file or grind the ends smooth. Place this piece of iron or steel, which should weigh ¼ pound (about 4½″ long) on end on top of one of the pats for an instant and remove it quickly. At the time when the bit of metal fails to make an indention, initial set may be said to have taken place, and the difference between this time and the time when the water was added, is the "time of initial set."

The final, or hard set, is also determined in the laboratory by means of a Vicat needle, but the student may take as the time of setting the period from the initial set to that time when the thumb fails to indent the surface of the pat. The student must bear in mind that the time of initial and final set is only approximate, as it depends upon the temperature and quantity of the water used in mixing and the air, also upon the humidity of the air, also upon the amount of mixing, molding or kneading which the paste receives. The amount of water is vitally important, and the following method may be used to determine the quantity required for any cement:

The Normal Consistency of any cement paste may be found very closely without the use of special instruments by mixing a ball of cement paste and then dropping it 20 inches upon a hard surface—a pane of glass will do—and noting whether the ball of paste flattens out or retains its shape and cracks in one or more places around the sides just above where it struck the hard surface.

Most brands of Portland cement require from 20% to 30% of water by weight, to make them into a paste of normal consistency. By paste, it is understood that the cement is mixed neat, i. e., using cement and water only. By mortar, it is understood that the cement has been mixed with more or less sand. By concrete, it is understood that the cement has been mixed with both coarse and fine material.

Weigh out a certain amount, say 10 grams of cement and mix it with water from a known weight of that fluid, say from 10 c. c. in a test tube. Mix the cement into a paste which appears to be about right, then roll it into a ball and drop it 20″ upon a level pane of glass. If the ball flattens, there is too much water. If it cracks, there is not water enough.

Assume that 3 c. c. of water was stirred into the cement as ascertained by weighing the test tube again after the paste was mixed. The ball flattens out as shown by A, Fig. 24, indicating that too much water was used in that particular ball. It will be well for the student to make a series of tests, say eight, and see how closely the results will conform to the results shown by Fig. 24, where each shape, A, B, C, etc., was taken from a ball which was mixed with a different percentage of water.

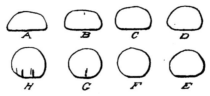

FIG. 24—NORMAL CONSISTENCY.

Thus, as we have found, ball A had 3 grams of water to 10 grams of cement, or 30% of water. The amount of water in each of the other balls varies 0.1 gram, or 1%.

The proportion of water in each test ball is shown by the following table, and the student should duplicate the weights in making a test for normal consistency. If the cement be the same as that from which these tests were made, then the balls will crack or "squat" with like quantities of water, but should the cement be of a different nature, even though quite as good, then the balls will crack or "squat" at percentages of water differing from those in the table. A table should be made up by the student for each kind of cement he has to use.

A—10 grams cement and 3.0 grams water = 30% Too soft
B—10 grams cement and 2.9 grams water = 29% Too soft
C—10 grams cement and 2.8 grams water = 28% Too soft
D—10 grams cement and 2.7 grams water = 27% Too soft
E—10 grams cement and 2.6 grams water = 26% Normal
F—10 grams cement and 2.5 grams water = 25% Normal
G—10 grams cement and 2.4 grams water = 24% Cracked
H—10 grams cement and 2.3 grams water = 23% Cracked

From the above table, it will be noted that balls E, and F, are about right, hence it is safe to say that 25.5% is just the quantity of water required for this particular lot of Portland cement. It will be interesting for the student to make a table of each cement he tests, and compare results. Referring again to Fig. 24, the balls E and F are seen to be pretty near their original spherical shape, and it may also be seen that while balls G, and H, are still better in shape, they could not stand the shock of dropping 20″ without cracking, hence they contain too little water.

The above table is also valuable in another direction. It shows that a fairly good ball can be made with water ranging from some point between D, and E, to a corresponding point between F, and G, and indicating that the water may vary between 24.5 and 26.5%, or the water in any neat cement paste may vary within 2% without much injury to the paste.

In making the 9 pats described before care should be taken that too much material is not mixed at one time. If all the material for the nine pats be mixed at one time, there will be trouble in keeping the mixture at the same consistency until they are all spread upon the glass plates. It may be better to mix only material for three pats at one time until after some dexterity is obtained in mixing pats and in spreading them upon the glass plates.

When making pats, number each one by scratching a number on top of each pat as shown by Fig. 23. This will allow each pat to be properly recorded and in noting the conditions of each pat there is less chance for error when they are thus plainly marked. By the time three pats are smoothed out on the glass plates, the first pat will be in condition to be marked readily and the marking can be done much better than the instant each pat is finished. The working of the surface brings out a film of water which gives the properly proportioned pat a sort of gloss after it has been finished. Until the surface water disappears, the marking is not as effective as it is after a few minutes have elapsed.

After the required number (9) of pats have been made, place them in moist air for 24 hours. It is well to arrange a vessel for keeping pats, test blocks, briquettes, etc., as there will be dozens of these shapes to be taken care of as the student reaches the testing of different concrete mixtures. . Almost any vessel will do, but the writer prefers a common sheet-iron baking pan 10″ x 16″ x 3½″ deep. Give the pan two or three coats of shellac and if signs of rust appear, dry out the inside and give it a couple more coats of shellac and there will be no more trouble from rust.

Put approximately ½″ of water in the bottom of the pan, lay some strips of board in the water and place the glasses carrying the pats on the strips of board. Cover the pan with two or three newspapers, place a piece of board on the papers, and the "moist air box" is complete and in working order. Examine the pats two or three times during the first 24 hours to see if any of the pats have cracked or swelled up in any way.

The pats should be smooth and fast to the glass at the end of 24 hours, with no change in any way. First-class cement will come out of the moist air in this way, but sometimes things happen, some of which are harmless, but some may be sufficient grounds for rejecting the cement. The several things which happen to pats are described in full on pages 96 and 97. The principal reason for using a pat thick in the middle, with thin edges, is that any expansion of the body of the cement will tend to enlarge the circumference, thus causing cracks along the edges of the pat.

Upon removal of the nine pats from moist air at the end of 24 hours, place three of the pats in water, to remain there 28 days. Place three more one side to remain in air for the same length of time, and place the remaining three pats in a kettle with bits of wood beneath and between the pats to prevent them or their glasses from touching each other. Fill the kettle with cold water to cover the pats, place on a stove and bring to a boil and continue the boiling for at least one hour. As the water evaporates, fill the kettle, to cover the pats, with boiling water from another kettle.

After boiling for one hour, remove the kettle from the fire, pour off the water and allow pats and kettle to cool sufficient to allow the pats to be handled. Examine each for any of the defects which can be perhaps shown in no better manner than by the accompanying sketches, in which are represented the following happenings, their cause, and what they represent:

Fig. 25—*a, b*—Shrinkage cracks, harmless—lack of care in making pats.

Fig. 25—*c, c, c*—Blotches, should be investigated—adulteration or underburning.

Fig. 25—*h, h, h*—Glass cracked, not dangerous—expansion.

Fig. 26—*d, g*—Pat left glass, dangerous if more than ¼″—expansion.

Fig. 26—*e, f*—Edges curled up, dangerous in water pats—expansion.

Fig. 27—Edges curled down, dangerous if more than ¼″—contraction.

Fig. 28—*i, i, i*—Radial cracks, dangerous, cause rejection—disintegration.

Fig. 28—*j, j, j*—Craze-cracks, cause rejection—complete disintegration.

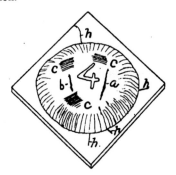

FIG. 25—SHRINKAGE CRACKS AND SURFACE BLOTCHES.

To describe more in detail what may be found about the pats when something is the matter with the cement, it may be stated that when you find little straight cracks like those shown by *a*, and *b*, Fig. 25, it means that the paste was either mixed too wet, or that it was dried out too quickly instead of being placed in air moist enough to keep the pat saturated until it had become fully set. Such cracks are harmless and only show lack of care in making up the pats and in looking after them when made.

The blotching shown by a difference in color at *c, c, c,* should be investigated chemically before accepting the cement. It may mean an adulteration of the cement, or underburning. Therefore, should the student find markings of this character in the neat cement pats, he will do well to use some other cement until—if he chooses—the cause of the markings can be investigated in a well-equipped laboratory and declared harmless or dangerous, as the case may be. Therefore, blotches may or may not be an indication of danger, and the cement can only be looked upon with suspicion until acquitted or condemned.

The cracking of the glass when the pat is sound, shown at *h, h, h,* is found only in water pats, as they will be called, to distinguish them from the pats left in air, or boiled. It is sup-

FIG. 26—EXPANSION CRACKS.

posed to be due to the expansion of the pat and the firm adhesion of it to the glass; and to the strength of the cement being greater than that of the glass. It is not usually regarded as dangerous. More expansion cracks are shown at *e,* and *f,* Fig. 26, but in the pat instead of in the glass. These cracks lie along the edges of the pat, not radial, and are very often found in air pats. In water pats, these cracks indicate too great expansive qualities in the cement and it should be rejected unless further tested and found to be all right.

The expansion cracks noted in the preceding paragraph are found when the edges curl up, with the body of the pat still adhering to the glass. Sometimes the pat comes off the glass and curves up so much that both sides are nearly alike instead of one side being flat. If the bottom of the pat 3″ in diameter curls up more than ¼″, it should be regarded as dangerous and rejected. If it happens in water pats (curls up at the edges and comes off the glass) it may be regarded as dangerous. This, however, very seldom happens in water.

FIG. 27—CONTRACTION.

Fig. 27 shows an occurrence exactly the reverse of that illustrated in Fig. 26. In this illustration (Fig. 27) the pat is shown curled down at the edges, the body of the pat having been

lifted clear of the glass. The same danger limit of ¼" of curvature applies to contraction of pats as well as to expansion.

Fig. 28 shows what is to be looked out for when testing a cement by the pat method. Radial cracks around the edges

FIG. 28—DISINTEGRATION

of a pat are always danger signals and no use should be made
of cement which shows cracks of this character. The cracks are
first observed radial, as at *i, i, i,* and in bad cases the cracks run
together like a "craze" in pottery, as shown at *j, j.* These cracks
should be looked for when the boiling test is made, and when
found, other tests, and the cement as well, may be abandoned at
once. When cement stands the boiling test, it is pretty apt to
withstand any other tests, though there are a very few cases
where the cement has passed the boiling test and has failed in
strength after several months. But the cases are so rare that
the boiling test may be accepted as an almost sure indication of
the quality of the cement.

It has been shown that a certain cement contained
from 38% to 40% of voids. It has been stated that there was
little possibility of filling voids in concrete by the addition of
cement, as that substance itself, contained nearly 40% voids, and

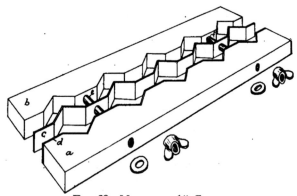

FIG. 29—MOLD FOR 1" CUBES.

it will be well for the student to make up some neat cement
blocks and test them for voids, specific gravity and weight to
the cubic foot.

With this outfit, there is also a block mold which will form at one time, five 1″ cubes. Another mold for 2″ cubes is furnished. The 1″ mold is represented by Fig. 29, the wing or thumb-nuts and the washers having been removed and the mold separated to show the manner in which the zinc lining is arranged. The main parts of the mold, a, and b, are held together by the bolts e, and f, which also pass through, and hold in place the zinc lining, c, and d.

When cubes are to be made, see that the mold is clean and well fitted together. Screw the nuts fairly tight with the fingers—there is no need of using a wrench on them. If the cube is to be tamped directly into the mold, the surfaces of the zinc should be well oiled to prevent the block or cube from sticking to the mold. It will, however, be found preferable to line the mold with paper before ramming the cubes. Paper prevents all sticking to the mold and it also allows the little cubes to be handled freely as soon as they are removed from the mold

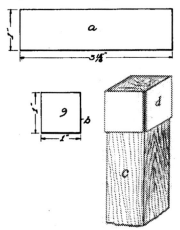

Fig. 30—Lining Cube Mold with Paper.

—something very handy, as the student will frequently find himself badly handicapped by having a lot of very tender cubes lying around, each in just the wrong place, and no way of moving them without knocking off some of the corners or otherwise disfiguring the little test blocks.

With a paper covered cube, there is absolutely none of the troubles met with above, and the paper may be readily peeled off after the cubes have been 24 hours in moist air, or the paper may be left on the cubes until they are to be weighed for the absorption test, when the paper is very easily removed by wetting (probably it is already wet) and then rubbing it off with a small scrubbing brush.

To line the mold with paper, cut up a lot of slips 1″ wide and 3 15/16″ long, as shown at *a*, Fig. 30. The slips are cut a trifle short of 4″ for the reason that they are much easier to handle when placing them in the mold. Procure a square stick which will slide easily into the mold; wrap the paper around stick *c*, as shown at *d*, making sure to start the paper at one corner and to rub down the folds snugly at each corner of the wood. Then remove paper from the square stick, place a numbered bit of square paper, *b*, in each space of the mold, press the squares down with the stick, then roll each creased piece of paper loosely around a finger and slip a piece into each space in the mold and twist the bits of paper around with the blade of a pen-knife until they lie square with the mold. Then it is ready for the material.

Mix 29.8 grams of cement with 10.2 grams of water and the mixture will weigh 40 grams. Ram this material into one of the cube-spaces of the 1″ mold. It should fill the space full, with a very little material left over. For ramming these little molds, nothing is better than a railroad spike cut off under the head with a hack saw and the end filed or ground very smooth and square with the body of the spike. Select a spike, if possible, which has a rounded point as shown at *b*, Fig. 31. The cut-off and squared end is shown at *a*.

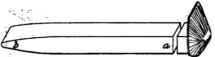

FIG. 31—A RAILROAD-SPIKE TAMPING-BAR.

An old saucer is just the thing for mixing material for one or two 1″ cubes, but for five, better use one of the agate dishes, preferably one about 8″ in diameter and 2″ deep, of which three should be obtained and kept for use in making mixtures and tests. Mix thoroughly with the pallet knife. An old teaspoon is an excellent tool for filling the mold with the saucer-mixed material. Drop a small spoonful in one of the molds and proceed to ram it with the wedge-end of the spike. This tends to crowd the material into the corners of the mold. In ramming any mold, look out for the corners. Do most of the ramming there, no matter whether making a 1″ cube or a 24″ block. Ram the sides and corners of the mold and the middle will take care of itself.

Ram lightly. Remember that the weight of the rammer is out of all proportion to the amount of material being tamped, therefore raise the spike a very short distance and let it fall on the material. Do not strike with the tool. Its weight is enough. Work around and around the mold and turn the tool so that the wedge point *b* lies in both directions occasionally. That is, do not let the tool cut the material in parallel lines all the time. Turn it one-quarter around occasionally so that the material is pressed to all sides of the mold.

The second portion of paste should be sufficient to fill the mold about half full. Work this as above described, turning the tool frequently and using end *b* most of the time. After the second layer of material has been well tamped with the wedge end of the tool, go over the surface with the square end, *a*, but be sure to go over the mold again with the end *b*, before putting in the next layer of material, which should fill the mold three-fourths full. Proceed as before with the wedge-end of the tool, only alternating with the square end, using this end more and more as the top of the mold is reached, the wedge-end of the tool being used exclusively at the bottom of the mold and none at all at the top, when the finishing is done by working all over the surface of the mold with the square end *a* of the tool.

After the mold has been filled and sufficiently tamped to compress the material as evenly and thoroughly as possible from bottom of the mold to the top, finish the cube with the pallet knife, smoothing surface of the paste and "striking" it off even with the top of the mold. Water should show itself on top of the cube when the tamping has been finished. Water should also show itself on top of each layer when tamped sufficiently. In case water does not show, the paste having been carefully mixed to normal consistency, it is evident that the layer has not been sufficiently tamped and tamping should be resumed until water is visible all over the surface of the layer. In ramming concrete, the same rule applies. Ram until water shows itself. If water will not "come up," the material is too dry. Care should be taken not to have too much water, permitting the large particles to work to the top of the mass, something which is sure to happen when we shake or stir a mass of various sized particles which are either very wet or very dry.

It is, then, very important that we use water enough in all mixtures to be tamped, and that they be tamped or rammed sufficiently. It also becomes apparent, from the floating of the larger particles, that when concrete or cement is mixed thin enough to pour, that it should hardly be tamped, as that process will only serve to bring the large pieces of the aggregate to the top of the mass. But, very luckily, as we shall find when studying mixtures of different sizes of sand and gravel, the larger portions cannot readily work to the top of the mixture unless there be too much fine material in the mixture. This is a very important matter, and it should be kept in mind.

The same law applies to neat cement paste. It is of little use to ram paste which has too little water in it. If there be too much water, the finer particles go to the bottom, the coarser come to the top. Therefore, when pouring cement paste, do not agitate it in any manner after pouring. And when ramming cement paste, make sure that it has water enough to bring it to normal consistency, then ram it until water appears on top of each layer of rammed material.

After the cube has been smoothed off on top, let it stand a few minutes if possible before removing from the mold, although when many test cubes have to be made up, it is not possible to allow them to remain in the molds after they are complete. But let them stay two or three minutes; if possible, ten minutes, then remove the thumb or wing nuts (Fig. 29) and separate the mold along the line between pieces b and c. The zinc strip will adhere to the cubes. Then the mold may be separated between d and a, the zinc strip d also adhering to the row of cubes. Then separate the strips of zinc, c and d, with the thumb and finger of one hand, while with the other hand holding the strips together about in the middle of their length. If the cubes stick to the mold, this treatment will cause them to separate from the zinc, which bends or springs under the pressure of the thumb and finger, peeling from the cubes without tearing them apart as might be the case were the strips of zinc suddenly forced apart along their entire length.

With paper around the test tubes, there will seldom be need of such extreme care in opening the mold, but where the paper is not used, it will be necessary to oil the mold every time it is used, also to separate the zinc very carefully in the manner described, in order to get a tender cube out of the mold without tearing off one or more of the corners. The mold should be cleaned nicely each time it is filled, and should never be closed for filling until every particle of adhering material has been carefully wiped from its surface.

As soon as a cube is removed from the mold, see that it is sound, with the corners intact and no holes or cavities visible at any part, top, bottom or sides. If paper was used, see that the strip was not smashed down during the tamping operation. Care should be taken that the paper is not thus driven down, also that no material is allowed to ge behind the paper. These blocks are so small that they must be as perfect as possible. Any little hole or poorly rammed spot reduces the volume or the strength of the small block in much greater proportion than when the same defect occurs in a large unit. Thus, it is important to make the curves as good as possible in order to secure all the strength in the tests to which we are entitled.

There will probably be some criticism about using 1″ cubes for ascertaining the strength, volume, etc., of the several mixtures. For physical tests in regular laboratories, 2″ cubes are used, also 6″ cubes, but for the needs of the student, the 1″ cube fills all requirements.

The use of the testing machine work is optional. For the use of those students who desire to send away to be tested, such cubes as they may make for this purpose, a mold for making 2″ blocks is sent out with the regular apparatus. This mold, as illustrated by Fig. 32, is shown with its bolts removed, and it is operated in much the same manner that the 1″ mold was handled. Paper lining may be used to great advantage in this mold also,

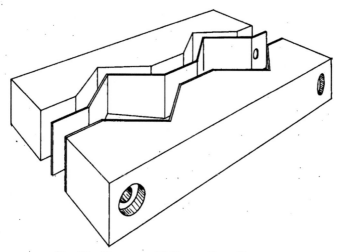

FIG. 32—MOLD FOR 2″ CUBES, BOLT REMOVED.

and the same railroad spike tamping bar may be used. The testing machine used in this work will take a 2″ cube, and will work up to 10,000 pounds pressure, or to 2,500 pounds to the square inch in a 2″ cube, thus the student procuring a testing machine may use 2″ cubes through the entire course if he so desires.

The 2″ cube is not necessary, for the reason that results obtained from a 1″ cube are always sure of being obtained from a 2″ cube. The reverse, however, is not true, as 2″ cubes often show a greater strength than 1″ cubes. Therefore, should there be any error, when testing out 1″ cubes, the error is always on the side of safety. That is, the larger mass of concrete will have more strength than the little 1″ cube. For this reason, the strength shown by the little 1″ cubes will never be greater than the strength of commercial blocks made from the same mixtures, hence we are safe in using the convenient 1″ cubes. These require much less labor and material than the 2″ cubes.

The cubes having been removed from the mold, see that they be plainly marked, then place for 24 hours in the moist air box. In this instance we will weigh the cube before it is put away, and we find it to balance the scale beam at 1.42 ounces = 40.25 grams. This weight must be carefully recorded to see if ‘the cube gains or loses in weight during subsequent operations.

As stated formerly, you should start a record book in which every measurement, every weighing, every action, and every test, should be recorded in so plain a manner that it can be found immediately, and read at a glance. Any blank book may be used for this purpose, but according to Benj. Franklin: "Whatever is worth doing at all, is worth doing well," and it will pay you to procure a loose-leaf book for this purpose. Such books may be obtained of any size required, but a set of covers to take "letter size" sheets is preferable. These sheets are 8½" x 11", and ordinary commercial letter paper is just the thing, holes being punched to fit the rings in the covers. Or, if desired, paper may be procured with loose-leaf covers.

If necessary, a bunch of loose letter-sheets and an ordinary box letter-filing case, makes a recording set not to be despised. The sheets should be numbered consecutively, and each one dated with the day, month and year. Put the date at the top of the page, the first mark you make on that sheet, then when you put it in filing case or book, add the page number and you can always get the page back into its place whenever it may have been removed from the file for reference.

The following record is from sheet No. 64 in the record book of tests made for this text, and it covers the neat cement block, the making of which has been described.

April 27, 1908   (64)

Mixed 50 gms. Portland Cement (brand) and 13 c. c. water,
Pressed part into a 1" cubical mold. After setting 1 hr. cube weighed 1.42 oz. (40.25 gms.).
Left on scale-pan over night.

April 28, 8:00 a. m.

Weight of cube = 1.34 oz. Loss = 0.08 oz.
9:30 a. m.
Placed to dry over range.

April 29, 9:30 a. m.

Cube weighed 1.224 oz. (34.62 gms.). Total loss, 0.196 oz. (16%).

Put into boiling water for 30 min., then into cold water until 11:00 a. m. Weight = 1.432 oz. (40.6 gms.). Gain = 0.208 oz. (5.9 gms.) = 17%.

Volume displacement = 17.27 c. c.

Volume absorption; $\dfrac{5.9}{17.27} = 34.1\%$.

Returned cube to water at 11:25 a. m.

April 30, 9:30 a. m.

Cube weighed 1.442 oz. (40.84 gms.).

1.442 oz.

1.224 oz.

0.218 oz. = Gain in weight.

$$\frac{0.218}{1.224} = 17.8\% = \text{Absorption by weight.}$$

Also 0.218 oz. = 6.18 gms., therefore—

$$\frac{6.18}{17.27} = 30.75\% = \text{Absorption by volume.}$$

May 6, 1908, 1:00 p. m.

Weighed 1.456 oz. or 41.2 grams, then $\dfrac{1.456 - 1.224}{1.224} = 18.9\%$

Volume absorption: $\dfrac{41.2 - 34.62}{17.27} = 38.1\%$ nearly.

You have before you, in this sheet, a complete record of the making and testing of a cement cube, and find its absorption gradually increases for about two weeks, over and above what it weighed when first made. It is also found that the absorption by volume is over 38%. It was ascertained that the percentage of voids in the cement was 38.9% before placing in the mold. As the voids have only decreased about 8/10 of 1%, it goes to show that cement does not lose any of its voids during the process of hardening or setting.

From the data contained in the record sheet (No. 64) the weight to the cubic foot of the neat cement block may be obtained. It has been stated that the specific gravity of any substance may be found by dividing its weight by its volume. In this instance the block occupies 17.27 c. c. of space, and weighs 34.62 grams. The specific gravity will be 34.62 ÷ 17.27 = 2.01. That is, the block is 2.01 times as heavy as water, consequently it weighs 2.01 × 62.57 = 125.5 pounds to the cubic foot.

In the calculations noted in the preceding paragraph, the volume, 17.27 c. c., was that of the cube when saturated with water, as will be explained in the next paragraphs. From the record sheet we find that the volume, 17.27 c. c., contained 5.9 grams or c. c. of water. Deducting that amount, the actual volume of the dried block, or the volume of the cement in that block or cube, is 17.27 — 5.9 = 11.37 c. c. Then 34.62 ÷ 11.37 = 3.05, the specific gravity, nearly equal to the original specific gravity (3.1) of the dry cement. But we must take the volume of the whole cube—not its net volume, with the voids counted out—therefore 125.5 pounds to the cubit foot may be taken as the weight of neat cement blocks made from the given sample of cement. When you try it, with a different cement, the result will necessarily vary a little from the figures here given, but the methods and processes are the same, and must be studied and worked until thoroughly understood, both in principle and practice.

In the record sheet is was stated that the

"Volume displacement = 17.27 c. c."

The next thing is the method of ascertaining the volume displacement. If the cube be exactly 1" square and 1" high, then its volume will be exactly 1 cubic inch, or about 16.4 c. c. But the block is not exactly 1" cubs. In fact, it is almost impossible to mold a cube exactly 1" on a side unless it be done in a very heavy mold and provision made to apply pressure and squeeze the cube down to exact size. But with tamped cubes, it is necessary to let the cubes go to any size they happen to take, and then determine that size as exactly as possible. The cube under discussion measures 2.55 and 2.65 cm. in one direction, at top and bottom of the cube, respectively. The other way measures 2.49 and 2.52 cm. The height, as measured on two opposite sides, is 2.64 and 2.66 cm., making an average of 2.6 × 2.505 × 2.65 cm. = 17.15. This comes pretty close to the actual volume of 17.27 c. c., but not close enough, and it is necessary to use some method instead of the measurement of the cubes to determine their volume.

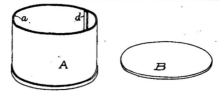

FIG. 33—VOLUME GAGE AND COVER.

The method used for volume determination is to weigh the volume of water displaced by the cube. This is done as illustrated by Fig. 35. The apparatus necessary is a Volume Gage, shown by Fig. 33, and consisting of the tin cup A, and the circular glass B. You can easily make these articles. The tin cup A is cut from a tin can, and should measure about 1⅝" in diameter by 1¼" deep. A Mennen toilet powder can makes a good cup if cut off and the edges ground very true all around.

The method of grinding the top edge of the cup is shown by Fig. 34. After the tin has been cut with tinner's snips as true as possible, then filed so true that it will not rock when laid upon a true surface—then it is ready for grinding. A sheet of fine emery cloth (sand paper will do, but it is not as good) is laid upon a smooth surface, the cup placed upside down upon the emery cloth as shown at a, and ground true by moving the cup sideways from b to c. The cup should be slowly revolved as

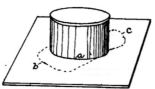

FIG. 34—GRINDING TOP OF VOLUME GAGE.

the grinding proceeds, and the grinding should be continued until no water will leak out of the cup when filled full, and the glass placed on top as shown by Fig. 35. The glass should fit so well that the cup may be turned upside down and not a drop of water leak out of it.

A very flat and smooth piece of glass should be selected. As shown at B, Fig. 33, the glass is originally circular in form, but a square piece would work very well, though it might not be as convenient as the round piece, which is easily cut a little larger than the cup. In fact, the cup is an excellent thing to cut around, the glazier's diamond being passed around the cup will secure about the required diameter of plate. If no diamond or wheel cutter is at hand, take an old pair of scissors, hold the glass under water and it can be readily cut to the required shape. Finish by smoothing the edges on an ordinary grindstone, using plenty of water to prevent heating and breaking the glass.

After the cup and glass fit each other, water-tight, place the cup in a dish of water, and when filled, with all air-bubbles out of the way, place the glass cover on top of the cup

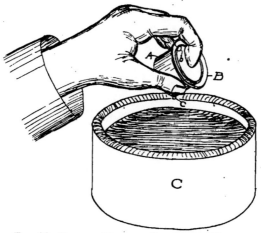

FIG. 35—FILLING VOLUME GAGE WITH WATER.

and lift out of the water as shown by Fig. 35, in which A represents the cup, held by the thumb and second and third fingers, while cover B is held firmly against the top of the cup by the forefinger. When the cup has been raised free from the water in dish C, let it drain a moment, and shake gently to dislodge from c any drop of water which might fall off, if left to itself. Tilt the cup as shown, in order that any water on top of cover B may run off. The idea is to get off all water which might run off under similar conditions. Take care that no bubble of air appears at b; should such a bubble show itself, return the gage to the dish of water and try again.

Having succeeded in filling the volume gage, with no air bubble visible, and no water adhering which will drop off,

FIG. 36—PLACING CUBE IN VOLUME GAGE.

place the gage in the weight-scoop of the scale and carefully remove the glass cover B from the gage, taking great care that no drop of water is spilled outside of the weight-scoop. It does not matter if water drops into the scoop, but take great care, yes, extra care, that no drop is lost.

Place glass cover B in the weight-scoop as shown by Fig. 36, in such a manner that it will not be in any water which runs over the top of A, and at the same time, no drop of water can get from A outside the weigh-scoop. Take the cube which is to have its volume determined and hold it between the thumb and forefinger, as shown by Fig. 36, and lower it slowly into the volume gage, taking care not to wet the fingers in the water inside gage, also not to drop the cube so as to make the water spatter out of the gage. The object of repeated caution in this direction is that, were a single drop of water to be lost, either by spatter- and it is very necessary that this water be not lost after filling away before it can be weighed, that amount of water causes the volume of the cube to appear just as much smaller than it really is, in proportion to the volume of water lost.

For instance, a cubic centimeter of water usually makes about 22 drops, as let fall from the small end of the pipette. As there are about 17 c. c. in one of the 1″ cubes, there would be $22 \times 17.27 = 380$ (nearly), therefore the loss of only four drops of water means that the result is in error about 1%. Accuracy is, then, very necessary, and care is necessary to secure accuracy.

Having placed the cube in the water-filled volume gage which in turn is resting in the weigh-scoop, carefully slide the glass B back over the top of A again, and see that no bubble of air remains underneath the glass. If the cup B stands fairly

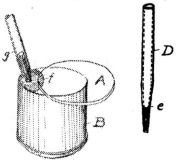

FIG. 37—DISPLACING AN AIR-BUBBLE.

level in the weigh-scoop, the cover may be repalced without catching an air-bubble under the glass. But if B does not stand level in the scoop, an air-bubble of more or less volume will be visible as at f, Fig. 37. The space occupied by the bubble must be filled with water before the operation can go any further, and recourse must be had to the pipette, to get rid of the bubble. Take in the pipette, some of the water in the weigh-scoop under gage B, and insert the small end of the instrument g, as shown at f, and let the water run out of g into B.

If the water does not readily run out of pipette g, place the lips over the other end of the tube and blow gently, whereupon the water will pass out in a hurry. In order to take up, in the tube, sufficient water to fill space f, it will be necessary to suck a little air out of the pipette, when the water will quickly fill the lower portion of the tube.

You probably have noted, when using the pipette, that a little water remains in the instrument, as shown at e, in sketch D. The bit of water is held in the tube by capillary attraction, and it is very necessary that this water be not lost after filling gage B. That water is wanted in the weigh-scoop in order to make the operation accurate. But it is almost impossible to get all the water out, no matter how much you blow into the pipette and dab the small end of the tube against the weigh-scoop—some of the water persists in staying in the pipette, therefore the best way is to allow that amount of water to stay in the pipette and not try to get it out.

Before using the pipette, suck an inch or so of water into it, from some source other than the weigh-scoop or volume cup, then blow the water out again, leaving the amount which persists in remaining after the tube has been used. This is usually about ¼″ to 5/16″ in the small portion of the tube, and if care be taken to see that this amount of water is in the pipette before performing the operation depicted in Fig. 37, then sufficient accuracy may be obtained and no water will be wasted in the pipette tube.

After gage A has been filled and cover B replaced without an air-bubble underneath, then raise the volume gage between the thumb and second and third fingers precisely as shown by Fig. 35, and again go through the same performance as described, getting rid of just as much outside water as described in that paragraph, in order that the water remaining on the outside of the volume gage when it is removed from the weigh-scoop may be as nearly as possible the same in amount as when the gage was placed in the scoop.,

If you perform the foregoing operations with care, the volume of water remaining in the weigh-scoop will be equal to the volume of the cube, which, it must be remembered, was full of water before being placed in the volume gage. A dry cube cannot be used in this way for volume measurement. It must be soaked as full of water as possible, and if in a hurry to make the test, boil the cube a few minutes to make sure that it has been heated clear through and all the water contained in it has been turned into steam. Then place the cube in water and as it cools off the steam will be condensed inside the cube, water will flow in to fill the partial vacuum thus formed, and only a very short time will be required to fill the cube to refusal, with water.

Weigh the water remaining in the scoop. In this instance it is 17.27 grams, showing that the volume of the cube is 17.27 c. c. Referring to page 105, you find that the weight of the cube dry was 24.62 grams, while after having been immersed in water a long time, it weighed 41.2 grams. The voids in the cube, therefore, are 41.2 — 34.62 = 6.58 c. c., and deducting that number from 17.27, the actual volume of the solid material in the cube is found to be 10.42 c. c. Thus, from the volume of the cube, its weight dry and wet, you can calculate the percentage of voids by weight and by volume, the weight to the cubic foot, and the specific gravity of the cube, or the specific gravity of the solids in the cube. Make up a set of five blocks as described in the foregoing paragraphs, note each upon a record sheet of its own, and calculate the specific gravity, absorption, voids, and weight to the cubic foot. Should you have a testing machine, test one cube when one week old, one at the end of 14 days, one at 28 days, keep one for your cabinet, and keep one for emergency in case a block gets damaged accidentally.

Watch each cube and when just hard enough to work well, scratch a number on for identification. To prevent the test numbers from becoming too large, a letter may be used for each cube in a batch, thus keeping the test numbers down. For instance, a certain test requires the making of 20 cubes—5 cubes each of 4 varieties—and to put all these cubes under one test number, which may be No. 3, call the blocks 3a, 3b, 3c, etc. Thus the test numbers will be kept small. But put some kind of a distinguishing number on every block or pat made, and record carefully and accurately, everything done, observed or assumed in connection with the test cubes. In this connection, it might be well to call attention to the decimal system, which has been successfully used for this purpose. That is, in series 3, say, number the cubes 3.0, 3.1, 3.2, 3.3, etc.

Take plenty of room on the record sheets. Nothing is so aggravating as to find some record so crowded as to be undistinguishable, or without room for an additional line if necessary. Any record may become of great value to you at any time, hence make it an unfailing point to enter data as neatly and accurately as you make the tests. And that should be as accurately and neatly as it is possible for you to make them.

CHAPTER VII.

MATERIALS OF CONSTRUCTION.

I—CONCRETE.

Applications of Cement:—The chief form in which cement finds its application is as concrete. Its main other applications are as mortar for masonry and plasters, both of which may with propriety be considered as corollary to the use as concrete. Indeed, mortar may be considered as fine concrete and ordinary concrete as coarse mortar, the same laws of making applying to both, but the usages differing. Hence a discussion of concrete in general will cover virtually the field. What is concrete? A material so well known and so widely applied would seem scarcely to need any special definition. Yet all is not concrete that may so appear; good concrete is one thing—poor concrete quite another. A brief statement, then, of what concrete is,— good concrete as distinguished from other kinds,—is in order.

Definition of Concrete:—Concrete may be defined as a mechanical mixture of loose solids, cement and water, in proportions calculated to produce a composite material having the least voids,—that is, a composite of maximum density,— which when first mixed is soft and plastic, but presently, owing to a chemical action taking place within and throughout the mass in which the cement and water figure, begins to solidify, eventually attaining great hardness and strength. The hardening process progresses continuously for several months, indeed for years, but in from a month to two months it reaches a practical maximum, when the structure may safely be put into use. The final product of this interesting process is a solid mass, limited · in continuity only by the compass of the structure, possessing unique strength, toughness, and resistance to the wrack and wear of both usage and time, capable of rendering efficient service indefinitely, if not for all time to come. Composed as it is of the hardest of materials, themselves among the commonest elements of the earth's crust and therefore everywhere available,—requiring only unskilled labor and the simplest appliances throughout the sequence of manufacture,—possessing perfect plasticity, and therefore fashionable with facility into any form right in position,—unequalled in durability and permanence,—efficient and economical from virtually every point of view, concrete, as a structural material, is, for very many purposes, without a compeer.

Reinforced Concrete:—Concrete, however, has one

fundamental weakness that would debar it from all usages where horizontal carrying capacity is required: which is its inability to stand stretching. Concrete of itself is so weak in tension that temperature changes even would disrupt it unless due allowance were made for this defect. It is for this reason that open or expansion joints are provided at intervals in all work embracing long stretches of massive concrete,—as for example a retaining wall. Most materials have the property of stretching—elongating, upon the application of sufficient pull, and if the pull is not so great as to strain the material beyond its elastic limit (point at which material, under tensile stresses, loses its power to regain its normal condition), it will, upon the release of the force, recover completely. Concrete, however, is unique in having no elastic limit, or, what is the same thing, its elastic limit and point of rupture are identical. Moreover, this point of rupture is comparatively low, ranging between 200 and 500 lb. to the square inch of section, depending on the quality of the concrete, and may run even lower. Therefore, it is neither safe nor economical to employ concrete in tension. The commonest example of material acting in tension is that of a string to a bow; similarly, in all horizontal structural members, such as beams and floors, the lower fibers are in tension, and should they yield the member will open up along the under side, and finally fail by doubling up jackknife-like, or breaking clean. Were it not, therefore, for the possibility and practicability of introducing some other material into concrete to be used for such purposes, adequate to supply this deficiency, concrete would be forever limited to massive work, such as foundations and walls. Haply, steel, a material strong and reliable in tension, admirably fulfills this need. Steel is supplied to the concrete in such amount and so disposed as to relieve the concrete of virtually all tensile stresses that may be expected in usage. The two materials together, being thus disposed so as to assume each the kind of stress for which it is naturally adapted, possesses the ability of acting to resist deformation as if they were one. To this composite material practice has given the name reinforced-concrete. It is also called Steel-concrete, Concrete-steel, Ferro-concrete, and Armored-concrete, and might, with most propriety, be called Structural-Concrete, which it really is.

The economy of the combination will be apparent: Assume, which is a fair average, that the cost of concrete is 20 cents per cu. ft., and of steel, 490×3 cents, or \$14.70 per cu. ft.; the cost of concrete is thus only 1/73 the cost of the steel per unit of volume. Assume also, which are the usual values, the tensile strength of concrete at 50 lb. and of steel at 16,000 lb. per sq. in., and the compressive strength of concrete at 500 lb., for steel being the same as the tensile. Then, for the resisting tension, 16,000/50, or 320 times as much concrete will be required as steel. Hence it will pay, by the amount 320/73, or a little over four times, to use steel for tension. But, for compression, only 16,000/500, or 32 times as much concrete will be required as steel, hence it will pay, by the amount 73/32, or a little over two times, to use concrete for compression. Thus, by using steel to assume all tensile stresses and concrete to assume all compressive stresses, the maximum economies of the two materials are developed, and a composite material produced which is cheaper than either one alone. This is the economy of structural-concrete.

History:—The use of cement for plain concrete is very ancient, although, like many other of the ancient processes, it remained comparatively a lost art until within recent times (see Chapter I), and has only attained its widespread, almost universal popularity through the discovery and development of Portland cement. The cement of the ancients was a natural cement, made mostly by grinding up lava, and was far inferior to our modern Portland cement; yet it was good enough so that of the structures of the ancients or parts of the structures, those of concrete alone have withstood all the ravages of the ages and remain today in good condition. The most notable example is the dome of Agrippa's Pantheon, in Rome, erected several centuries before the Christian Era, which is today in perfect condition after nearly 25 centuries of use and exposure to the elements. The Roman Forum and Appian Way, too, abound with examples of ancient construction. With these testaments before us of the permanence of concrete, users of the material today may well rest secure as to the life and usefulness of the present-day works.

Reinforced-concrete, however, is a modern discovery. First used supposedly by Monier, a Frenchman, for making pottery, along in the early seventies, its development now, in the early part of the 20th century, is still in its infancy. Indeed, its broad application to the purposes of constructional engineering is a development of the last decade. It is hard to overestimate the far-reaching influence the ingress of this material has had upon the building world. Its effect has been indeed revolutionary. It has extended the application of cement to include almost every conceivable kind of structure; more than that, it has created, is creating, and will continue to create, new forms of usefulness,—forms, indeed, before the advent of this unique and versatile material, not even vestured by fancy. Declared a prominent architect recently, "Architects see with this material (reinforced-concrete) the realization of their fondest ideals of design;" and Thomas Edison, the wizard of the electrical world, in a recent interview (March, 1909) prophesied: "Within the next 20 or 30 years—and it will start within the next two or three—concrete architecture will take enormous strides forward; the art of molding concrete will be perfected, and what is equally important, cheapened; there will rise up a large number of gifted architects, and through their efforts cities and towns will spring up beside which Turner's picture of ancient Rome and Carthage will pale into nothingness and the buildings of the Columbian Exposition appear common. But great expense will not attend this; it will be done so that the poor will be able to enjoy houses more beautiful than the rich now aspire to, and the man earning $1.50 a day, with a family to support, will be better housed than ·the man of today who is earning $10."

Principles Involved in Combination of Steel and Concrete:—The possibility and value of the combination of the two materials, concrete and steel, seemingly so opposite in every respect, to form virtually a new material possessing the strong point of each and the weak points of neither, comes ·about on account of the following facts:

(1) That to all practical intents and purposes the co-efficients of expansion of the two materials are identical, so that under a varying temperature they change length at the same rate; (2) that the concrete, in changing from fluid to solid, shrinks just enough to engage the imbedded steel bars in a grip from which under proper conditions it can only be released by the stretching and therefore drawing-out of the metal; (3) that steel encased in concrete is immune to its common agencies of ruin, rust and fire; (4) that concrete properly reinforced with steel will work with the steel to resist tension without yielding until the yield-point of the steel itself has been reached, and then the cracking is confined not to any particular region, but is well distributed along the length of the imbedded steel bars. Thus it becomes possible, by suitably reinforcing concrete, to apply it to any and all situations where bending and transverse, as well as crushing stresses, and also temperature stresses, are occurrent. This obviously has enormously enlarged the sphere of usefulness of the material,—in fact has brought within its compass practically every conceivable kind of engineering structure,—more than that, it has made possible entirely new and original types of construction.

WHAT A PERFECT CONCRETE WOULD BE.

Conception of a Perfect Concrete:--A perfect concrete is theoretically but not practically possible; yet a consideration of the same is useful in setting before one a picture of the ideal to be striven for. A perfect concrete would be one in which there were no voids or pores, but the different grades of materials entering into its composition would be so nicely sized and intimately mixed as to eliminate all interstices or spaces between the different granules. Such a concrete would possess great density, hardness, toughness and strength, and moreover, would be impervious to fluids even under pressure. If a number of grades of perfectly spherical flinty particles could be obtained,—ranging in size say from an inch down to the infinitesimal granule, even impalpable powder, and in mathematically correct amounts, so that the voids between the largest balls would be neatly filled by a group of the next smaller diameter, and the voids between these in turn similarly filled by still smaller balls, and so forth, and if a perfect grouping of these particles were also possible, the result would be a perfect concrete. The accompanying figure represents such a conception. The finest particles, which would be fine enough to slip into the smallest conceivable· spaces and thus virtually coat every particle coarser than themselves, would be the cement. Combining chemically with the water these would form an unbroken crystalline glaze, coating every particle of the mass, filling every space not elsewhere

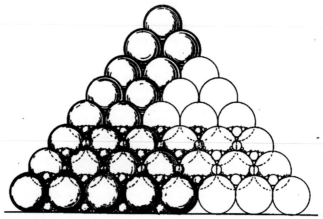

This Diagram represents Stack of Balls, with interfeces packed with Balls of a regularly decreasing Diameter, in an attempt to eliminate the Voids in the mass. This is the principle of the Concrete Mix; in proportion as realized is density & strength of concrete, and efficiency of a given amount of cement: the more perfectly voids are filled the less cement relatively is required.

FIG. 38—SKETCH SHOWING A GRADED CONCRETE, USING SPHERICAL INERTS.

filled, and interlacing and interlocking the whole into one unbroken mass.

Value of Conception of Perfect Concrete:—Such a mixture is of course out of the question, nor is it desirable; but indicating an ideal it is useful as an object in thought to hold us up to as close an approximation as possible. In the end it will be found in this, as in all things, that efforts made on the part of the *user* tending to greater accuracy will be amply repaid in character of results obtained and reputation established. The best way may not always be the cheapest in the long run, but it very generally is; and this is one of the cases where it is.

Approximations of Practice:—The mixtures used in practice at the present time are at the best but crude approximations to what a concrete ought to be, and it is safe to say that they must and will be improved as knowledge and skill in the are increase and become more wide spread. The future will not tolerate slovenly and haphazard methods nearly so much as the past has and present does, but to less and less extent. The worker in cement today who would rank as a master of the art tomorrow must *today* thoroughly master the essentials and strive everlastingly for increasing proficiency in their application.

HOW CLOSE WE MAY COME IN PRACTICE TO A PERFECT CONCRETE.

Limiting Conditions:—There are many considerations that prevent us making as good a concrete as we would like. In the first place it is difficult to get always materials of the proper quality, and in the second if they could in any instance be obtained—sized and graded after the manner outlined in the preceding paragraph—it would be next to impossible, with the appliances at hand, to mix them perfectly, and dealing with human nature to always be assured of proper workmanship. *But what we can and must do is this: Ascertain the percentage of voids in each lot of inerts and endeavor by supplying a sufficient amount of a finer material to reduce these voids to a minimum.* This may require the admixture of several grades of material. But whatever the requirements, the results obtained in increased efficiency of the concrete and in actual saving in amount of cement required,—said amount decreasing directly as the voids,—will be found to justify any and all efficient expenditure of time and money to that end. In the long run it will be found that the *better balanced the mix, not only the stronger the concrete, but the more economical.* Falk, in his book "Cements, Mortars and Concrete," records an instance where with the same materials a mixture proportioned one part of cement to nine parts of graded sand and broken stone, gave better results on testing than one proportioned one to six, and concluded that it is because the former was better balanced. In this case improving the concrete was coincident with cheapening it. Leonard B. Wason, of the Aberthaw Construction Co., of Boston, in CONCRETE ENGINEERING, May, 1909, states similarly: "The saving in cement alone more than paid for the extra cost and trouble of sizing and testing the aggregates," referring to results obtained on one job where a sizing of the aggregates into five lots had been made and the proper recombination of these determined carefully by experiment. *Thus there is no excuse for tolerating indifferent and indiscriminate mixtures.* And, also, we can and must see to it throughout the sequence of mixing and placing until the concrete is in its final position that the details be faithfully executed. In reinforced-concrete work especially is it absolutely necessary that the quality of the concrete be as good as possible—and it is with reinforced-concrete that we are for the most part herewith comcerned.

Current Mixtures:—It has been customary in times past to designate the composition of a concrete mixture as so many parts by volume of cement to so many parts of sand to so many parts broken stone or gravel. For instance, a 1-2-4 mixture means one part cement to two parts sand to four parts coarse aggregate, measurement being ordinarily by loose volumes. Common mixtures, as so characterized, are:

1-1½-3, very rich, used for columns and piers, etc., where very high compressive strength is required; 1-2-4, known as a rich mixture, commonly used for reinforced-concrete, etc., and for all purposes where a dense, impermeable concrete is required; 1-2½-5, a medium rich mixture, also used for reinforced-concrete work, especially massive portions, like footings and thick walls; 1-3-6, an average mixture, used for massive plain concrete, like walls, foundations, buttresses, arches, etc. If used for concrete that is to show a rich facing mixture of stone, chips and cement are usually combined with it, being placed at the same time; or surface may be "plaster" afterwards; 1-3-7, a lean mixture, used for massive concrete under ground, where surface voids are not considered objectionable.

Mixtures as lean as 1-3-9, 1-4-12 and 1-6-12 are also used, but they are excessively stony and have a very limited application, mostly for filling purposes, where a porous concrete having some coherency is desired.

These proportions have been adopted by practice tentatively insomuch as it has been found that in general they give the desired results. Perhaps the best balanced mix would be the second or third, although it is probable in both these more large aggregate could be used without impairing the strength and density. But examples are numerous and multiplying that these or any other stated proportions should by no means be religiously adhered to; instead, they may be misleading, and strict adherence to them, as is often blindly insisted upon by some overly zealous inspectors, may defeat the very object intended. A mere statement of proportions is no guarantee of quality. The best concrete can only be obtained by considera-
a of each particular case along the lines already indicated,
ns they will be hereinafter elaborated.

"So far better way of expressing the proportions would be: similar v parts aggregate to one part cement, measured under mediate nditions, and the maximum, minimum and inter-
izes of aggregate stated and the proportions of

each size," qualified by the clause—"that the mixture shall be so graded by selecting the aggregates as to secure the maximum reduction of voids, and then cement supplied in amount sufficient to fill the remaining voids with, say, 10 per cent excess."

There is much need for reform of existing practice in this respect.

Common Errors:—Common errors that work to prevent the attainment of a good concrete are these: Stress is laid upon the question of *quality*, while that of *quantity* is indifferently heeded or ignored, and vice versa; emphasis is put upon the *cement* while all other components are practically ignored; mixtures are made to conform to arbitrary standards regardless of the common-sense requirements of the matter. The result is variously an indifferent concrete or a very poor one; occasionally by happy chance a good one.

Relative Importance of Ingredients:—The *cement* is in a sense the most important element that enters into concrete construction, for it is virtually the life of the combination, and in reinforced-concrete is the chief factor in the basis of the interaction of the two materials. The other elements, although forming the bulk of the mass and very important factors in the strength and durability of it, are inert,—that is, loose and nonadhesive, and until impregnated with cement are so much dust. However, *it is only by the proper intermingling of the proper amounts of the proper materials that the mixture may be transformed from a loose, non-coherent, practically worthless lot of dust, into a homogeneous solid of structural value.*

All Elements Important:—The point it is desired to emphasize is this: *That no one element is of all importance and all others inconsequent.* In the most majestic bridge the apparently insignificant little pin is as vital to the integrity of the structure as the largest and most imposing chord. Should it be absent or fail to perform its relied-upon duty the safety of the entire structure may be imperiled. So in the making of concrete—plain or reinforced, substantial or ornamental, hidden or exposed, foundation or superstructure,—from pit to pinnacle, it is imperative that every element from the minutest grain to the largest pebble, from the thinnest strand to the thickest bar of steel, should be right in quality and amount, should be thoroughly and properly incorporated with the rest of the ingredients, and the whole carefully and correctly placed in position for the finished structure. Not until in its final resting place and sufficiently solidified to be able to stand for and by itself, should vigilance be relaxed. So concrete, while seeming perhaps to the casual observer and the would-be user absurdly simple to make and mold, is in reality one of the most difficult materials to apply, for *with no other structural material are such scrupulous*

*care and unceasing vigil prime essentials to success. What chiefly
determines a good concrete is the amount of intelligence used in its
making.*

WHAT ARE THE PRIME QUALITIES OF A GOOD CONCRETE?

The **chief requisites** of a good concrete are that it
should be dense, hard and strong, tough and non-vibrant; durable;
constant in strength; unaffected by fire and fumes; impervious
to water; immune to the agencies of decay.

Quality of Ingredients:—Naturally enough, this requires that the ingredients of concrete—the aggregates as they are
commonly called—be selected with due regard to these same
considerations. *A good concrete cannot result from the blending of poor materials, however perfect the blending be.* All
materials of an unstable nature, in a state of or prone to decay,
fragile or brittle, acted upon by acids or acid fumes, melting or
shattering under the action of moderate high temperatures such
as might be expected in an ordinary conflagration, unduly porous
or spongy, containing impurities of an organic or acidic nature,
etc., are obviously unsuited to the purpose. A discussion of
various concrete materials with reference to these qualifications
will be given in subsequent pages.

Quality of Blending:—Naturally enough, also, this
requires, for a good concrete, that the blending be properly accomplished; *an indifferent blending of the best materials,* even
though present in proper relative amounts, *cannot produce a
good concrete* any more than a perfect blending of unsatisfactory
materials.

Quality—Complete:—The whole is only the sum of its
parts; if any part is lacking the sum is imperfect and the whole is
not a whole at all. Good concrete results from, and only from, a
proper blending of proper materials of proper size in proper
proportions.

FUNCTION OF THE INGREDIENTS OR CONSTITUENTS OF CONCRETE:

Foreword:—In the foregoing paragraphs the subject
of concrete has been treated collectively; now it will be considered with relation to each of its parts. The function of all
of the parts taken together is to produce good concrete—the
best practicable. The function of each one of the parts is to
perform its individual duty so perfectly as to detract nothing
from the efficiency of the whole. The various ingredients or
constituents of concrete will therefore be considered in this
light, beginning with the cement, which may, and rightly, be
regarded as the most important element.

THE PART PLAYED BY THE CEMENT.

Primary Functions:—The *cement* has been called the *life* of the concrete; it is the element that binds all parts into a homogenous whole. By itself—dry, it is useless. Reduced to a thin paste by admixture with water it becomes capable of performing its *duty as a binder*. The office of the water is two-fold: (1) To liquify the cement and thus put it in shape to coat the rest of the ingredients; (2) to make possible the crystallization of the cement. The cement also performs another function. It constitutes *the finest filler,* making the final reduction of voids, virtually filling every space or interstice not already occupied. That concrete is the densest and strongest in which the minimum of voids remain to be filled by the neat or clear cement,—for the reason that cement in chunks is both pervious to moisture and fragile, hence unsatisfactory as a filler. It is also the most economical, inasmuch as the less voids to be filled by cement,—that is, the less cement acts as a filler and more as a pure binder,—the less the amount of cement required to produce a given result, hence the more economical the mixture.

Practical Functions:—*Practically the cement, as mixed with water, performs another very useful function, namely that of a lubricant,* facilitating the process of mixing and placing. It acts in effect, by "greasing" the various inert particles, to convert the whole into a heavy liquid. In this connection it is interesting to compare the former practice with regard to the means and manner of mixing concrete with the present accepted method,—that is, the so-called "dry" concrete with the "wet" or sloppy concrete. Then it was thought that the amount of water present in the mix, that is the amount of liquid cement, should be a very minimum, and that density should be secured by a great deal of vigorous pounding. Now it is known that enough water, that is liquid cement, should be present to make the mass flow easily, virtually eliminating tamping in the sense formerly understood, the result being not only a marked reduction in the cost of mixing and handling, but actually a denser and stronger concrete. It is logical and economical to mix concrete "wet." Another advantage of the fluid concrete, which rightly deserves mention here, is that by this very property the sphere of usefulness of concrete has been very widely extended. It has made possible the effective and economical molding of concrete in the widest conceivable variety of forms and positions, and the attainment of artistic surface textures. All this is due to the lubricating action of the cement cream. It is this that allows the various particles of the combination, during the process of mixing, handling and placing, to slip and slide on one another until, under the influence of gravity, assisted by mechanical agitation for the purpose of eliminating air bubbles and the congestion of particles, they assume a position of maximum stability,—a condition coincident with correct positioning of all particles and therefore a mixture of maximum density and strength. It is this, also, that makes possible and practicable the attainment of a smooth, hard, dense and impervious surface finish, of uniform color and texture, integral with the body of the mass for, by using smooth, truly formed material for molds, a slight amount of spading of the freshly deposited material next to the face of the mold, suffices to yield a very satisfactory surface finish. Finally, but not leastly, reinforced-concrete, by far the most important and extensive application of cement-concrete, would never have progressed out of the chaos of experiment into an actual, practical material had it not been for the introduction and development of the "wet" mix; the *liquid-cement* not only makes it possible and practicable to properly encase the reinforcement, but is indispensable to the development of an adequate bond or union between the two materials. This question of bond is so important that it will be discussed separately.

Summary of Principal Functions:—So it is seen that the *cement,* in addition to performing its *essential function as a binder,* serves also *several important auxiliary functions,* and that it becomes capable of performing these various functions, primary and secondary, only by virtue of being used correctly. How its various essential properties affect or govern its value or efficacy in their several respects, and hence measure its usefulness, will be subsequently discussed.

How the Cement Concerns Fireproofness of the Concrete:—The *cement* also enters directly into the question of *fireproofness*— one of the most important qualifications sought for in a building material. The exposed surfaces of concrete structures are or should be smoothly coated, all voids neatly filled and all coarse aggregates covered by a film of cement mortar, such being drawn next to the smooth surface of the mold by capillary attraction during the operation of placing and tamping. This film being mostly cement and sand is naturally proof to wide extremes of heat and cold. The heat of a conflagration, however, may be sufficient to drive off the water of crystallization of the hardened cement and thus resolve it back into loose form. Such action ordinarily takes place between 800 and 1,000 degrees F. But even in the case of continued exposure to such a temperature for many hours the disintegration seldom would exceed ¼ to ¾-in., and the heat penetration to the inside be inconsequent. Concrete is naturally a poor conductor of heat. Coated with a film of dehydrated cement it becomes practically nonconductive. Thus it comes about, on account of the amount of heat required to dehydrate set cement and evaporate the released moisture, and the extraordinary low conductivity of dehydrated material, that cement concrete possesses unique fire-resisting properties, ranking all other materials. However, the degree of fireproofness depends on the nature and quality of the inert ingredients, which is a topic for separate discussion, but it is plain that the part played by the cement, and as present in the proper consistency, is of prime importance. The condition of proper consistency is insistent, since if the mix be too dry, the exterior surfaces can hardly present the same unbroken glaze of cement to the attack, with the result that the effect will be more serious. Nor can the defect of a rough surface, occasioned by improper consistency, be corrected by a plaster-coat, as the latter, not being integral with the body, would be certain to peel off at a comparatively moderate temperature, leaving the rough surface to the freer attack of the flame.

Much of the low conductivity of concrete and the fact that the steel remains cool throughout a fire-test is due, no doubt, to the chill produced by evaporation. It is the phenomenon of evaporation of any fluid that cold is produced, and the more rapid the evaporation the greater the degree of coldness. This fact is made use of in processes of refrigeration.

THE QUESTION OF BOND.

The Chief Element of Bond:—No discussion as to the *function of the cement* would be complete without due consideration of the part played by the cement directly as well as indirectly, in making possible the intimate inter-action of the two materials, concrete and steel, by which the material known as reinforced-concrete, or more properly *structural-concrete*, comes into existence. The grip of the concrete on the imbedded steel members is commonly known as the *bond*. The amount or efficiency of the bond is influenced by several considerations, chief of which is the gripping action induced by the shrinkage of the concrete as it hardens. All concrete in air contracts slightly as it hardens, such contraction extending possibly over several years, but coming to a sensible limit within three to four months. This amounts in all perhaps to from 1/5 to 2/5 of 1 per cent of the volume,—varying with the proportion of cement. The richer in cement the greater the relative shrinkage. Under water, on the contrary, the opposite phenomenon is observed, namely that a slight swelling in volume is experienced, which also varies in amount directly with the richness of the mix, but is relatively only about half as great as the air shrinkage. In the one case the grip is normally sufficient, so that a bar of plain steel imbedded for about 50 diameters will pull apart before it will pull out; in the other case it may be so insufficient as to call for a special reinforcement, equipped with adequate anchorage.

The value, obviously, of the gripping action, vice-like as it is, is influenced by the smoothness and evenness of the contact of the gripping medium, the concrete, about the periphery of the engripped medium, the steel,—the more nearly perfect the better the bond. Here is where the cement in fluid form comes into direct action. Drawn by capillary attraction to the smooth steel it tends to coat same as a varnish, thus bringing about intimate and continuous contact between the concrete and the

Diagram illustrating states and stages in the immersion of a round (cylindrical) solid in a heavy fluid, e.g. mercury or linseed oil.
(1) Surface film just beginning to yield and stretch under pressure of solid.
(2) " " stretched to breaking point—solid half immersed.
(3) " " snapped and joined again overtop of solid just immersed.

FIG. 39—DETAIL SHOWING A ROUND ROD IN CONCRETE.

steel for the full length and full section of the latter element. It also, by inducing fluidity of the entire mass, results—by the very property of a fluid—in the soft material flowing flush all around each member of the steel element. It thus gives the uniform contact around the periphery of each, so essential to the development of an adequate bond and thorough protection to the steel against the agencies of ruin. The round section of

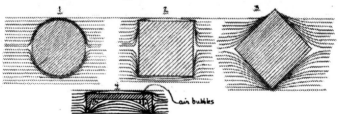

air bubbles

Diagram indicating manner in which surface film of fluid tends to arrange itself about various structural shapes:
(1) Cylindrical section; contact smooth and nearly perfect.
(2)+(3) Square or diamond section; contact varies, being best at corners.
(4) Channel section; contact least perfect, with tendency to collect air bubbles.
Evidently, the particles of fluids tend to adapt themselves most perfectly to round shapes.

FIG. 40—DETAIL SHOWING ACTION OF CONCRETE AROUND
VARIOUS SHAPES OF STEEL.

steel bar, affording as it does a uniform curvature of contact, is obviously best suited for the full and proper development of this element of bond; an attempt to demonstrate this fact has been made on the diagram accompanying (See Fig. 40). Experiments, too, bear out this contention. Other reasons favoring the use of round sections will be presented in connection with the consideration of the reinforcement.

Friction Bond:—*Another element of bond is formed by the mechanical inter-locking of the surfaces of contact.* On first thought a bar of plain steel as it comes from the rolls, seems perfectly smooth; yet not nearly so smooth as if turned down and polished in a lathe; its smoothness is actually relative, not very pronounced after all; viewed with a magnifying glass it would be seen to be not smooth at all, but fairly rough, as represented diagramatically, as in Figure 41; indeed, to a vision fine enough to perceive the minute particles of cement,

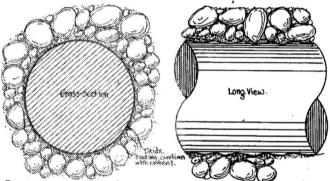

Diagram showing manner in which concrete encases steel bar. Roughness somewhat exaggerated to show interlocking. Cement seen to be next the steel.

Fig. 41—Enlarged Detail Showing the Action of Concrete on the Surface of a Steel Rod.

almost impalpable powder, as large as pin-heads, the surface of an ordinary bar of plain steel would appear as rough as a rip-saw. Hence, before slippage of the imbedded steel can ensue, it is necessary for all the innumerable little nodules of cement and cement-mortar engaging the surface of the steel to be sheared off. It is to this agency and the friction-contact due to shrinkage that the bond chiefly owes its value; and we have seen how essential the cement is to both.

Deficiency of Dry-Concrete and Origin of "Deformed-Bar:"—In this connection it is pertinent to observe that one of the most marked deficiencies of "dry" concrete, as used in the form of structural-concrete, is the inadequacy of the bond developed; a condition brought about by the lack of sufficient "juice" (cement-cream) in the mixture, without which *contraction* of the mass cannot take place, nor can a satisfactory coating of the steel with a film of cement integral with the rest of the mass ensue—both of which have been shown to be indispensable to the development of an adequate bond. It was in the days of "dry" concrete that *"mechanical-bond"* bars were devised to meet an apparent need; but under present conditions the bond realized with plain bars is, with one exception,—namely sub-aqueous construction, usually entirely adequate.

BOND BETWEEN STEEL AND CONCRETE.

Type of bar.	Age in days.	Size of specimen.	Lbs. per sq. in.	Authorities.
¼ round Plain................	60	6 in. diam.	315	1904 tests University of Illinois. 1:2:4 limestone.
¼ square Plain................	60	6 in. diam.	325	
¼ Johnson..........	60	6 in. diam.	477	
¼ twisted lug.......	28	8 x 8	511	F. H. Shibley, at C. S. A. S., 1907. 1:2½:5 stone.
¼ Ransome	28	8 x 8	401	
Thacher.............	28	8 x 8	441	
¼ round Plain................	60		264	A. S. for T. M. 1:2:4 limestone.

Chemical Adhesion:—*There is yet another element of the bond* which, while by no means so important as those already mentioned, is nevertheless a factor of value: This is *the chemical adhesion of the cement to the steel.* This action is supposed to be based upon the chemical interaction of the cement with the film of oxidized iron always found coating steel that has been exposed to the atmosphere,—which may be present in sufficient quantity to be visible as a reddish film, or in advanced stages as a crust, or it may be invisible to the naked eye. So long as the oxidation has not progressed to the acute stage, when it becomes a source of weakness and is not to be tolerated, it is not harmful, but rather to be desired and for the reason that by the chemical action above signified the value of the bond is increased. There seems to be, between this oxide coating and some property of the cement, a chemical affinity resulting in the formation of a new compound,—supposedly a limy hydrate of iron,—which makes for a closer union of the two materials. Just how great is this effect it is, of course, difficult to determine, especially since it varies with the amount of oxide present, but experiments seem to indicate that it may increase the total bond value in excess of 25 per cent. At any rate, the fact that what oxidation has already taken place when the steel is used is effectively neutralized and further oxidation checked forever, did it not contribute an iota to the bond, would yet be an important factor in the art, protecting' and preserving as it does the imbedded metal from corrosion. Several theories have been advanced to explain the immunity of steel encased in concrete from corrosion, that is, rusting. One is that it is due to the imperviousness of the concrete, which is not always a fact; another that the cement, which is of course, strongly alkaline, neutralizes whatever acid there may infilter with air or moisture, thus removing from the latter the elements to which corrosion is supposedly due; still another, and the one intimated above, that the cement forms with the iron-oxide 'a glassy protective coat which prevents acid-laden particles from coming into direct contact with the raw steel. Probably the net result is due to the combined action of these three; at all events the presence of the *cement*, and as a *liquid*, is seen to be instrumental in each and all.

Summary of Elements of Bond:—Summarizing, the bond of concrete and steel is due to:

(1) Friction-contact or gripping brought about by the contraction of the mass,—a change in volume made possible by the property of the cement in setting.

(2) Mechanical interlocking of the surfaces of contact; brought about by the fineness and fluidity of the liquid-cement.

(3) Chemical interlocking of the surfaces of contact; results from chemical interaction of the cement with the oxide-coating of the steel.

In all these it has been seen that the *cement, present* as a *cream* or *thick fluid,* is the *vital instrumentality.*

MISCELLANEOUS PARTS PLAYED BY THE CEMENT.

Durability and Permanence:—The cement directly concerns the question of *durability* inasmuch as the extraordinary resistance of Portland cement concrete to abrasion and the action of the weather and time is largely due to the extreme hardness and chemical stability, of the cement itself. Concretes of inferior or unsound cement yield easily to wear and tear and soon crumble, while good cement is so resistant that artificial grind-- stones in which it is the binder will outlast natural grindstones.

Waterproofness:—The *cement* also contributes largely to the *waterproofness* of concrete, filling the voids and discouraging the passage of water through the mass, if not denying water ingress by reason of its hard. smooth surface. Instancing the last, a common way of waterproofing concrete surfaces is to brush them carefully with a paint of neat cement and water, or again by plastering the surface intended to hold water with a rich cement mortar, say one part cement to one part fine sand. Masses of pure or neat cement are not the most watertight, as might off-hand be supposed, the porosity, on account of the uniformity of the grains, being relatively greater than in ordinary concrete. Moreover, there is always a considerable quantity of air entrapped in cement which is hard to work out—and this increases the porosity. Consider: The specific gravity of cement is about 3—that is a solid cubic foot of cement will weigh from 187 to 200 lb., whereas a bag of cement, measuring loose 1 cu. ft., will weigh only 95 lb.; hence, in loose cement there is approximately 50 per cent voids. The presence of this air is attested when a quantity of cement is poured into a bucket of water, bubbles arising like the working of a ferment. Concrete, however, by carefully mixing proper amounts of properly graded materials, and by thoroughly kneading the fluid mass in the molds to work out the bulk of the air, can be made dense enough to resist successfully water pressure. So, in making water-tight concrete, these things are to be scrupulously observed if success is to be realized. The question of waterproofness is an important one and will be given treatment separately.

Neutral Properties:—Cement is strongly *alkaline* in its reaction; instancing this is its effect on the bare fingers of the worker in it, being similar in its action to potash-lye, and requiring the use of gloves, preferably of rubber. In the concrete mixture this property gives a valuable purpose in *neutralizing* or *counteracting the effect of acids.* Air and water nearly always are laden with acid elements, the products of nature and of industry; these would tend to erode the concrete and especially corrode the metal reinforcement, were it not for the counteractive effect of the cement. Even the rocky mountain finally yields to the attack of such agencies and crumbles to dust, whence indeed has come into existence in the course of ages the soil that mantles the earth's crust,—the interior, below the surface-soil—all is solid still. But cement-concrete, combining the most resistant of materials and possessing inherent protection against the common agencies of destruction, is well calculated to out-endure even the hills themselves. It is to this quality of the cement that imbedded metal owes its immunity to corrosion. The pioneer Ransome once subjected a small block of concrete, through the center of which a steel bar extended, to the action of the salt-sea for a period of several years; on removing it he found the projecting ends of the metal rusted to a shred, but upon splitting open the block found the metal inside perfectly intact. Others have made similar experiments with like results. The concrete was, however, of *proper material* and *mixed wet.* With porous or spongy aggregates, as for example common cisders, and with dry-mixed concretes, the *steel* is *not perfectly protected,* and for the obvious reason that air and water may pass through to contact with the metal without necessarily coming in contact with any cement. Whenever steel has been found to oxidize in concrete, the cause has been traceable to one or more of the following: (1) Porosity of the aggregates, (2) porosity and permeability of the concrete, (3) dryness of the mix, and (4) the development, owing to improper design, of cracks, destroying the continuity of the protective coating and exposing the metal to ready attack. All these are avoidable, and to-be-avoided, faults, and if avoided there will be no trouble with the steel rusting.

Interaction With Iron:—The *protective* effect of *cement* on *steel* is also supposed to be due to the formation by the cement, or elements of the cement, with the incipient film of iron-oxide always forming on naked iron, of a vitreous compound of iron, which, forming a thin coat around the metal effectually guards it from contact with acid-laden moisture or air. Mention of this property is made in the paragraphs on *Bond,* and as there pointed out it is dependent upon the cement being used *wet*.

Effect of Electricity:—*Cement* is a non-conductor to the electric current; but steel is a good conductor and, moreover, rapidly corrodes under the action of the electric current. Here again cement comes into office as a protector of the steel against disintegration by electrolysis, a danger attendant in many structures or parts of structures due to stray or escaping electric currents. And once again it is requisite for the success of the cement in this case that it be used *wet*; for otherwise it could not be present either on the exterior of the concrete or along the imbedded steel in a uniform, unbroken coat, as is obviously necessary.

Preservation of Steel in Concrete:—In regard to the preservation of steel in concrete the following witness by Mr. H. C. Turner, of the Turner Construction Company, builders in reinforced-concrete, is of interest and of value *(Engineering News,* Jan. 16, 1908):

"The Turner company recently had occasion to wreck a reinforced-concrete one-story structure, erected by this company in 1902, at Brighton, L. I. The building was on a pile foundation, the piles being cut off at mean tide level; footings, columns, side walls, floors and roof, all of reinforced-concrete * * * All steel reinforcement was found in perfect preservation, excepting in a few cases where the hoops were allowed to come closer than $\frac{1}{2}$-in. to the surface, some evidence of corrosion being found in such cases, thus demonstrating the necessity of keeping the steel at least $\frac{3}{4}$-in. from the surface. The footings were covered by tide-water daily, yet the concrete was in all cases extremely hard, and showed no signs of weakness from the action of the salt water. The steel bars in the footings were perfectly preserved, even in cases where the concrete protection was only $\frac{3}{4}$-in. thick."

Professor Norton, of the Massachusetts Institute of Technology, after an exhaustive series of experiments to determine this point, stated among his conclusions:

1. "Neat Portland cement is a very effective preventive against rusting.

2. "Concrete, to be effective in preventing rust, should be dense and without voids or cracks. It should be mixed wet when applied to steel.

3. "It is very important that the steel be clean when imbedded in concrete."

As another instance, William Sooy Smith, M. Am. Soc. C. E., states: "Upon removing a bed of concrete at a light-house in the Straits of Mackinac, 20 years after laid, and 10 ft. below water level, imbedded drift-bolts were found free from rust."

Again, in Bulletin No. 35, U. S. Department of Agriculture, by Allerton S. Cushman, on the "Preservation of Iron and Steel," the following is stated:

"The question whether steel imbedded in concrete is protected from corrosion is of the highest importance. If steel reinforcement rusts away it bodes ill for the future of many modern structures, owing to the impossibility of making inspections and repairs before the danger point is reached. The record of discussions before a number of engineering and scientific bodies show that there is at the present time a conflicting opinion in regard to this subject. There can be no doubt that the reaction of unleached cement is strongly alkaline, owing to the separation of free lime at the time of set. If this alkaline reaction is maintained, steel imbedded in concrete should remain uncorroded. If, however, as is sometimes the case, percolating waters find their way through the concrete the free lime will eventually be removed and dangerous rusting take place. Nails and other objects of steel will remain bright in lime water."

In commenting upon these statements and findings, the vital importance of a dense, impermeable, and sufficiently reinforced concrete is apparent.

THE PART PLAYED BY THE WATER.

Chief Functions:—The *function* of the *water* in the concrete mixture has already been pretty well indicated. Concisely summarized, its office is two-fold: To impart *fluidity* to the entire mass, enabling the cement to perform its various duties, and facilitating the operations of mixing and placing; secondly, to enter into chemical combination with the cement, bringing about the phenomenon evidence in the hardening of the mass by which in due season a strong, dense, solid structure results,—that is, the *crystallization* of the cement.

Quantity Required:—The *quantity* of water required

for a given mass of materials is governed by the first considera-
iton rather than by the second, inasmuch as any deficiency in the
amount of water as may be required for proper crystallization
may be subsequently supplied for the purpose of set, as was the
case with "dry" concretes. Sufficient water for fluidity assures
enough for every other purpose. The water should produce
a mass soft enough to flow freely into position and yet not separate. If there is a slight excess, over and above what is necessary for proper consistency, it will rise to the surface and be
evaporated. To preserve a uniformity of conditions, interior and
exterior, the surface of new concrete work should be kept wet
by suitable means for as long a period as practicable—the
longer the better; for evaporation may remove the moisture
from the surface particles so rapidly as to set up non-uniformity of
"set," injurious to the mass as a whole, and, moreover, often
ruinous to the immediate surface.

Quality Required:—As to the *quality* of the water,
it should be fresh, clean, free from acid, oils, and organic matter in suspension or solution, and preferably "soft,"—that is, free
from mineral matter in solution. Sea water possesses salts in
solution that are deleterious and so should not be used. These
are salts possessing an affinity for the aluminous element in the
cement that may constitute, by reason of the formation of expansive compounds of alumina, a distinct danger; especially would
this be likely to occur if the cement used happened to be high
in *alumina*. Mention was made of this contingency in Chapter III,
under *Alumina*.

THE PART PLAYED BY AGGREGATES.

Chief Function:—The *aggregates* comprise the bulk
of the mass; their function is to provide "body" to the material.
In the general discussion preceding, the part played by the aggregates has already been so clearly indicated that little elab-
oration is needed here. The chief point to be noted is the better
the grading of the aggregates the denser and stronger the concrete, the less the amount of cement required for a given amount
of inert materials, and the more economical the mix.

Classification of Aggregates:—*Aggregates* are commonly divided into two classes, the one called *fine* and the other
coarse; this is a purely conventional separation. In practice
this division is evidenced in the use and separate specification
of two classes of materials, sands and crushed rock of small
diameter being classified as *fine,* and gravel and broken stone as
coarse. By a proper combination of these two, the *fine* filling
the voids in the *coarses,* a dense body for the concrete may be
formed. In each case the proper proportioning of aggregates
should be the occasion for careful experiment, and if the common

materials at hand fail, as they will often do, to yield the required density, the deficiency must be supplied by importation of suitable amounts of the size or sizes lacking.

Effect on Density:—The *density* of the concrete depends upon the aggregates, both as regards their proper sizing and their individual quality; soft or porous aggregates obviously cannot make dense concrete.

Effect on Strength:—The *strength* also depends on the proper grouping and the proper quality of each and every size of aggregate; the strength of the whole is, in a degree, only the strength of each individual particle, and like a phalanx, is dependent on intimate association of all units.

Durability:—The *life* of the concrete, too, is dependent upon the aggregates; materials soft, fragile, brittle, easily acted upon by the agencies of decay, and poor resistants to fire and gases, cannot make enduring concrete, no matter how well mixed,—the weaker the individual particle the weaker the mass as a whole. Often perfectly satisfactory materials are unobtainable except at very great expense, and then it may be necessary to compromise on the most satisfactory; it is for such a reason that limestone, although a poor resistent to acid and fire, yet possessing many other good qualities, finds use in concrete.

Size of Material:—The question of *maximum size* of aggregates is governed rather by the practical consideration of the size of structural members and reinforcement than by the requirement of strength or density of the mixture; obviously, the maximum size should not be more than such as to pass freely into the forms and between the various pieces of steel. In large masses, as large stone as is practicable to handle may be, and from the standpoint of economy, should be employed; but ordinarily such should not be considered as one of the constituents of the concrete-mix proper, rather they are independent foreign bodies introduced to swell the bulk. The maximum size is really determined by the consideration of fluidity, particles so large as not to merge readily with the fluid-mass, rather to tend to separate from it under handling, being obviously out of place; if used at all they should be added to the mixture as it is placed in the molds and deeply imbedded, acting to swell the bulk as above indicated.

Shape of particles:—The question of *shape* is more theoretical than practical; theoretically, as outlined hereinbefore, the spherical shape is ideal; practically, while aggregates of rounded gravel and sand make a little stronger and denser concrete, irregular, jagged particles, like broken stone, give satisfaction, indeed may be preferable on account of superior quality.

THE PART PLAYED BY THE STEEL REINFORCEMENT.

Chief Function:—The *steel reinforcement,* or more properly speaking, the *steel component* of *structural-concrete,* is virtually the *backbone* of the material; it is that which imparts stiffness and coherency to the whole, enabling it to bridge, carry and maintain an unbroken structure; it is, to use a homely parallel, to the engineering structure what the skeleton is to the body. Wherever in the structure, under the action of external forces or loads, there are set up internal stresses tending to separate the material,—that is, tensile or stretching stresses, there steely-fibers in the shape of steel bars of small diameter, are introduced; these, by virtue of the intimate contact or bond of the two materials, are enabled to interact perfectly with the concrete, acting to assume the tensile stresses while the concrete assumes the opposed or compressive stresses, thus maintaining the balance or equilibrium of forces necessary to stability; also, effectually preserved by the encompassing medium, the life of the steel becomes co-existent with that of the concrete. The correct amount and proper positioning of the steel within and relative to the concrete is, in any case, subject to calculation, and is taken up in detail in works on design. The question of *quality* of the steel will not be considered here.

General Function:—The *general function* of *the steel* as a *tie,* lacing, as it were, all parts of a concrete structure together, is important in more respects than may, on first thought, appear. Concrete is notoriously weak in tension, especially when green, and it is then that cracks may occur upon removal of shores that would not if the material were sufficiently aged; this requires that steel be used in greater quantity and in different positioning than the bare theory of working stresses would seem to require *There should be a continuity of steel throughout the structure in every direction, and especially span-wise, extending uninterruptedly over all supports;* then, if over the supports the reinforcement be elevated, thus simulating the catenary curve, the concrete may be considered as resting on the steel as a saddle, providing alignment and protection for the steel, and a stiff floor. In this case, then, the integrity of the structure is virtually independent of the bond. As it costs but very little if any more to place steel in such a manner, and as by so doing it becomes possible safely to remove the shores and put the structure into use the sooner, it seems a justifiable measure, especially as the bond does not begin to become a dependable quantity until after several weeks. These are questions of such vital concern that they will be given detailed consideration by themselves.

Temperature Stresses:—*Another important function of the steel is to assume the stresses induced in the concrete structure by the changes of temperature.* All materials change length, either shorten or lengthen, with the shifting temperature.

It has been pointed out, as one of the fundamentals at the basis of the interaction of steel and concrete, that practically, the co-efficients of the two materials are identical. This co-efficient, as fairly definitely established for steel, is 65/10,000,000, or decimally 0.0000065; that is, for each degree* Centrigrade change in temperature the length changes that ratio; thus a bridge 1,000 ft. long at 10 degrees C. would increase in length for each degree rise above 10 degrees, 1000×0.0000065, or 0.0065 ft., which is about 5/64 in. Steel stretches and lengthens as the temperature grows hotter and hotter, finally liquifying; concrete, on the other hand, has little elasticity under the effect of temperature, and is a very poor conductor, so that a comparatively small degree of change causes it to pull apart or crack and not to soften and stretch. It is to allow for this that monolithic walls of concrete unreinforced are provided with open joints every 30 or 40 ft. Concrete reinforced with steel in sufficient quantities, however, acquires the property of transmitting temperature strains without cracking. This new property comes about through the instant transfer of the stretch by means of the bond, into the steel, and the stretching of the concrete being thus uniformly distributed throughout the mass no visible cracking results until the steel itself begins to yield. Minute cracks may appear before the steel begins to yield, and probably do, but they are so widely and uniformly distributed and so fine as to be of little consequence. The amount of steel required to assume temperature stresses may be readily calculated; usually in a reinforced concrete structure, such as a building, little or no attention is paid to the effect of temperature in the lines of direct stresses, the reinforcement provided for the regular working stresses being considered sufficient to care for this also.

Surface Cracking:—*Other offices the steel* may be called upon to perform are to prevent cracking of the surface under the action of fire or the heat of the sun, for which fine rods or wire mesh imbedded close to the exposed surface are much used; the prevention of serious cracks or disjointing in case the structure happened to be subjected to unusual conditions, such as unequal settlement of the foundations, the effect of seismic disturbances, or impact of falling bodies. In each case the steel acts as a binder, assuming the incident stresses and holding the mass together.

Compression:—*Steel may also act in compression,* the bond in this case acting to transfer the load from the concrete to the imbedded steel; this use, however, is more of an expedient and discussion of it will be reserved for books on Design.

---

*The Centigrade thermometer is the common standard for all scientific determinations; zero begins at the freezing point of water, corresponding to 32° Fahr., and 100 is at the boiling point, corresponding to 212° Fahr.; a degree C. thus is 9/5 a degree F.

SOME SPECIFIC FAULTS IN CONCRETE WORK.
DISCUSSED.

Classification of faults:—*Faults* in concrete work are of *two kinds:* Those due to the use of *bad material;* those due to *bad workmanship.* The worst condition results when these two faults combine. Then little short of a miracle will save from disaster. The evils of both have already been pointed out in general, but here some of the more frequent specific faults will be discussed.

Quantity Faults, the Cement:—If there is *too little cement, everything else right,* the mixture will be stiff and "creaky" in mixing, difficult to place, the attainment of a smooth, hard surface improbable, and the resulting concrete weak, rough, and porous; if *too much cement,* the mixing and placing will not be materially influenced but the surfaces of the finished work will dust and peel and the whole mass tend to crack from excessive shrinkage.

Fine Aggregates:—If *too much fine aggregate* is used, the mixture will appear mushy instead of fluid, the finished surfaces sandy and easily gouged or scraped, and the strength of course more or less seriously impaired—an excess of fine particles is perhaps the most serious weakness in concrete, "eating up" the cement and thus sapping the life of the mixture. *Too little fine stuff,* on the contrary, means a hard mixing concrete, "rocky" in appearance, difficult to place, and generally porous and pitted, owing to the insufficiency of mortar to fill the voids, also likely to be fragile. It is like a pop-corn ball.

Coarse Aggregates:—*Too much coarse material* in the mix increases the difficulty of mixing, impedes the flow, tends to destroy fluidity, and to separate out from the rest, giving a "stony" concrete, rough on the surface and liable to pockets of large aggregate devoid of fine material,—alike unsightly and detrimental; *too little coarse material* rather eases the mixing and placing but it means a "slushy" mixture, lacking in body and in toughness.

Excessively Sized Aggregates:—If there are any *sizes too large* they will not mix easily with the rest and will tend to separate from the body under handling, as mentioned in the paragraphs on *Aggregates;* but, as also pointed out there, they do no harm if added to the mass separately and care is taken to have them neatly imbedded; the one necessity is to see that they are clean and saturated with water so as not to deprive the cement by absorption of the moisture necessary to its set,—then they will incorporate thoroughly and add rather than detract from the strength. In massive work especially is such material permissible and desirable, effecting often a marked economy.

The Water:—If there is *too little water* the mixture

will be stiff and dry, be difficult to place, and require excessive tamping to get results, possessing according to the degree of dryness, weaknesses as previously noted; *if too much water,* the mixture will not hold together,—the cement will tend to leach out and the large aggregates to segregate and scatter, more or less seriously, endangering the integrity of the concrete and giving an inferior surface. A slight excess of water, not enough to cause appreciable separation and leaching away of the cement, is comparatively harmless; it will rise to the surface and there stand until removed by evaporation or the hand of man.

Detection of Quantity-Faults:—In *general,* a *deficiency* or *excess* of any ingredient may be detected by the appearance and behavior of the mass as it comes from the mixing; the more nearly everything is right the more perfect the fluidity of the mixture, the easier to mix, handle and place, and of course the best and cheapest from every point of view. The trained judgment can thus gauge the fitness of concrete, as far as proper balancing and mingling of the mix is concerned, by the simple appearance. If the mixture after placing and tamping appears smooth and dense, with no visible voids, it is a rough indication that the mix is proper.

Workmanship-Faults:—Suppose the *materials* in *quality* and *amount* are *satisfactory,* and that the success of the mixture depends entirely on the method and manner, the care and thoroughness, with which the blending and placing is done. Then common faults that may manifest themselves are: Segregation of fine particles here and coarse particles there; clotting of the cement; congestion of coarse material resulting in considerable voids; pockets and voids intermittently distributed; porosity; fragility; low strength; indifferent to worthless concrete, spoiling much good material. About the worst fault of this kind, if there can be any distinction drawn, is the clotting of the cement in the form of balls of dry material or simply pieces of pure cement. In either case a double source of weakness if introduced; cement, by itself being porous and fragile, its congestion means "rotten" spots in the concrete; and, being congested instead of distributed, it means that the entire mass is deprived of its proper vitality.

Combination of Faults:—Proportionately as any combination of the above faults manifest themselves, and they are one and all due to ignorance and unfaithfulness, is the mixture an indifferent success or a failure, and the efficiency and very life of the imbedded metal and therefore of the entire structure, endangered. See paragraphs on the *Parts Played by the Cement.*

Quality-Faults:—The final supposition is that *everything is right except the quality of the materials* themselves. It is common sense that concrete is only as good as the materials out of which it is made and that inferiority of any one, such as for instance bad cement, is liable to seriously impair the whole if not ruin it; "a chain is only as strong as its weakest link" applies with full force here. The qualities required of materials for concrete have already been generally indicated, and they wil) be discussed more specifically in articles to follow.

Faults Concerned With Reinforcement, Concrete Responsible:—*Faults due to or in connection with the steel component* may be classed under two heads: (1) Those concerned with the placing of steel; (2) those concerned with the placing of the concrete. Normal conditions ensue when the steel members are properly positioned in the molds and secured there prior to pouring the concrete; then the burden of fault will be entirely upon the concrete. Let the material be too stiff or dry and the steel will not be properly covered, nor will the conditions be right for the development of the bond; let it be deficient in cement and the bond will suffer also; let the aggregates be over-sized,—that is, having pieces too large to pack around and between the steel and there will be trouble due to the arching of coarse fragments, resulting in large voids in the region of the stone, therefore inadequate adhesion and protection of the steel. Let it be too stiff, and if the section is other than round, the material is quite likely not to flush up on the other side of the steel, resulting in deficient bond, pro-. tection, and hollow spaces underneath the steel in the bottom of structural members. Let it be too wet and the proper amount of fine stuff will not flow around the bars, but rather flow away, leaving only a mass of coarse particles, slimy with cement-water, to encompass the steel, obviously a bad condition. *It is essential that the steel be uniformly imbedded and any feature of the concrete mix that tends to defeat this object is manifestly bad.*

Steel Responsible:—The burden falls upon or is shared by the steel under the following conditions: (1) When it is not placed prior to but along with the concreting; (2) when the prior positioning is insecure so that displacement takes place during the concreting; (3) when an attempt is made to crowd or pack the bars too closely. (There is generally a multiplicity of bars making up the steel component of the member.) The custom of placing the steel, or attempting to, along with the concreting is a product of crude experimental conditions, when any auxiliary devices such as would be required to position the steel in the molds prior to pouring the concrete were regarded as an expensive luxury—if indeed there was at that stage in the development of the art enough forethought to have conceived of such. In some sections of the country yet today loose-bar installations, as they are commonly called, still find vogue, but haphazard and slovenly methods are unjustifiable and must give way to more scientific conditions. Foremen seem unable to appreciate the importance of exact, not approximate, placement of the steel. It is pertinent to observe, as a sign of the times, that the oldest, best trained, and most skilled constructors voluntarily adopt such measures as may be necessary to insure the exact, not approximate, position of the steel in the finished structure. Results of not taking such precautions are: (1) Congestion of bars in bunches, not allowing proper amount of concrete to encase each bar; crowding of a bar or bars to one side of the mold, resulting in slight or no protection to the steel and insufficient bond; (2) positioning of one or more bars either too low or high, in the one case giving insufficient protection and bond, and in the other reducing the efficiency of the steel itself, since the closer to the neutral axis the bars may lie, the less they come into action; (3) uneven bedding so that while one end may be all right the other may be far too high; (4) bars too far one way falling short of one support and overlapping the other an unnecessary amount; (5) mistakes in amount and size of steel, as called for by the design, attended by even more serious results—unless steel is positioned before attempting to concrete it cannot be checked either in amount, correctness of sizes, or position. Needless to add, all these are grave faults,—the *steel must be positioned properly as called for by the design, and suitably secured and carefully checked, before attempting to pour concrete.* The third case noted,—to wit, when an insufficient space is allowed for each bar, is a matter of intelligent design. A competent designer who is not carried away by penurious considerations or an unholy desire to skimp, will not fall into this error. It means insufficient bond, inadequate protection, great difficulty in placing the concrete so that it will flush in all around the bars, and thus a serious weakening of the entire member.

SOME COMMON OBJECTIONS TO CONCRETE DISCUSSED.

General Origin:—It is safe to state that most of the objections registered against concrete are occasioned, not by any inherent shortcoming of the material itself, but rather by its misuse and abuse. It is to be remembered that the industry is still very young and that its growth has not been slow and gradual but mushroom-like—unparalleled in the rapidity of its development in the history of industry. Hence errors were to be expected. But we are passing rapidly into a healthy maturity—order is emerging out of chaos; intelligence is supplanting ignorance; system haphazard; and as a result objections are fast disappearing. Concrete—king of structural materials—is coming into its own.

Chargeable to Dry-Mixing:—Many complaints against concrete may be chalked up against the *fallacy of dry mixing;* that it is rough and uncouth, fit only for underground work;

that it is not impervious to moisture; that it is not a fireproof material (see discussion of part played by cement in making concrete fireproof); that the adhesion between concrete and plain steel is inadequate (see article on cement and bond); that it is forever limited to massive work; that it admits of little or no architectural treatment; that it is difficult and costly to mold; etc. We have seen how with the fluid-mix all these and similiar objections are swept away, until now only the mists of former prejudice remain.

Bad Workmanship:—Many objections to concrete will be found upon careful investigation to be rooted in the results of ignorant and careless workmanship. Most of the extended failures, as well as the minor ones, may be traced to this source. But the experience of concrete in this respect is only the universal experience with new things, and as the art becomes more firmly established the occasion for complaint on this score must and will pass away. A discussion of faults of this kind has been already given. (See preceding article.)

Bad Design:—Another class of objections group themselves about the *question of design*—especially as concerns the *reinforcement*. Practically all cracking of parts of concrete structures both at an early date and subsequently under heavy usage fall under this *head*. Early designers, failing to recognize the monolithic nature of the material, and thinking of it in terms of wood, brick, and steel rather' then in terms of itself, neglected to provide for many stresses with steel reinforcement. Some of these stresses—peculiar to the material—were fundamental and some unimportant. The fundamental ones were those existing over and approaching the supports, induced by the monolithic nature of the material. It was not sufficiently realized that with concrete there is no longer simple, isolated, disjointed action from support to support—as with timber and steel—but that action is continuous throughout the structure, and that if the design were not adapted to this continuity of action, tensile stresses were going to be thrown upon the material in a place and manner with which it (un-reinforced) could not cope. The result was the development of cracks—some serious and all unsightly and undesirable—over and in the neighborhood of the supports. Many of the early structures (constructed 10 or 15 years ago, and even quite recently) evince faults of this kind, although nearly all are still carrying heavy loads. And several of the failures of concrete buildings within recent years, happily during the construction stage, were largely due to this lack. Notably the Bridgeman building in Philadelphia, which collapsed in 1907, and in which failure—in addition to utter lack of provision for stresses over the supports—there were manifested virtually all the faults peculiar to this class of work—*the fruits, one and all, of ignorance and willful neglect.* The principle is simple; a con-

crete structure in its action is analogous to a suspension bridge,
—none would think of erecting such without an absolute tie
across the towers or supports, insuring continuity of action between the shore and river spans and securely anchoring all ends
of cables or suspension rods where continuity ceased. It is
very much the same in concrete. The steel reinforcement performs a duty akin to that of the suspension cables.

Incomplete Analysis:—Faults due to neglect of *minor stresses* are concerned mainly with the action of temperature, eccentric loading, settlement, wind, and earthquake.
A concrete structure should be a unit throughout, equally strong
in every part for the purpose intended, and provision made for
all probable stresses—ordinary and extra-ordinary, incidental
as well as primary. There should be an adequate tie of steel
everywhere and in every direction. Then, defects of this kind
will cease to appear—provided of course the materials are all
right and correctly applied.

Surface Defects:—*It is claimed that concrete surfaces
craze or check, and often discolor and effloresce,* all of which
defects have, and are liable to, occur; but they are one and all
due to improper cement or the improper use of a good cement.
Concrete containing soluble salts or organic matter, due to the
use of dirty or impure aggregates, will discolor; and if made
with cement containing unsound or unincorporated elements, it
will effloresce—if not swell and crack, and possibly disintegrate.
Concrete surfaces too rich in cement may check under the action
of temperature and sun rays, and even ordinarily rich surfaces
often do likewise.

Dusting of Surfaces—is due for the most part, to (1)
use of material too dry, (2) disturbance of surface after initial
set has begun, (3) neglect to keep surface wetted for a sufficient
period afterward. The last mentioned is perhaps the most common
cause of "dusting", and for an obvious reason—that the moisture is removed by evaporation from the surface particles of
cement before they have had chance to properly hydrate, and
unhydrated cement is still loose powder. The "dusting" of
wearing surfaces has been overcome by oiling the surface, or
treating it with a specially prepared compound, or rubbing it
down with a carborundum brick. Exterior surfaces have been
treated by tooling or acid-etching (which eats away the loose
particles of cement); and of interior surfaces by painting or
whitecoating.* Dry, rough, and porous molds are also a cause
of "dusting", this is the disadvantage of using wood-molds.
However, oiled, and then well wetted immediately prior to

*Perhaps the best paint for interior surfaces is white Portland Cement itself, applied
mixed with cold water by dashing with a brush to the surface previously well wetted
and kept wetted subsequently for several days. This can be applied with hose and
spray pump.

placing concrete, wood will be little trouble on this score. The reason is plain; absorbent molds deprive the surface particles of the cement of their proper moisture, producing the same effect as noted above,—that is, non-hydrated cement on the surface. Thus, the best molds are the most non-absorbent; attesting this, concrete cast in metal or damp sand molds does not "dust."

Inartistic:—That concrete admits of little architectural treatment. This objection has been largely met by the introduction of the fluid-mix, for with a plastic material of this nature there is practically no limit to the intricacy of detail that can be worked out; almost any form that can be molded in metal or carved out of stone can be herein reproduced with facility and economy. The matter of a pleasing surface texture is being met by the exercise of care in the selection of suitable aggregates for surface material, and the after treatment of the surface in some manner that discloses the color ot the aggregates and imparts a light-diffusing quality.

Ponderousness:—That concrete is ponderous and unsightly,—objections due sometimes to "finicky" or amateurish design, but more often a matter of taste. The substantial appearance of most concrete structures is rather an advantage, and in comparison with the old brick and stone types the proportions are indeed slender. For a real fireproof structure it is difficult to conceive of any other existing form of construction more compact and dignified than *structural-concrete*. In tall structures, where adherence to a strict structural-concrete design would give piers of inconvenient girth, the difficulty may be met nicely by a judicious combination with structural steel,—keeping down the girth of piers to a minimum by coring with structural steel while using the ordinary construction for the floor system. In this manner there is almost no limit to the height of concrete structures economically and structurally obtainable. The tallest concrete building in the world at the present time is the 17-story Ingalls building in Cincinnati, an office building of neat and trim appearance, in which there is not a pound of structural steel.

Resistance to Fire:—*That concrete is not fireproof,*—a question already partially discussed. If concrete, as now constructed, is not fireproof, then no material at hand is fireproof. Concrete has been subjected to some very severe tests—actual as well as experimental, and in every *bona fide* instance it has shown an extraordinary high resistance, meeting the endorsement of the experts upon careful investigation after the two greatest catastrophes of recent years, the fire at Baltimore (1904) and the earthquake and fire at San Francisco (1906). And it has been demonstrated time and again that concrete floors, composed of floor slabs and beams, after subjection to the ultra-severe conditions of a test—far more severe than any actual trial—still retain a considerable portion of their original carrying capacity, and upon the disintegrated material being removed and the surface brought to form and alignment by plastering, are almost as good as ever for service. Concretes made of hard-coal cinders especially exhibit extraordinary high fire resistance, but cinders have other disadvantages that limit their use. Rock-concretes are sufficiently resistant for all ordinary purposes.

Mr. Richard L. Humphrey, in charge of the government testing laboratories at St. Louis, in which the relative merits of materials of construction are tested, had this to say, in summing up his conclusions as to the showing of the San Francisco fire and earthquake:

"Brick columns gave fair satisfaction, but concrete-protected columns afforded the best results.

"The writer is of the opinion that the present commercial hollow terra cotta tile is largely, if not entirely, devoid of merit for fireproofing purposes. Even when it is of the best grade it can hardly be considered a first class building material.

"Concrete, especially reinforced, because of its great adhesive strength and its reinforcing metal, proved more satisfactory than any other material. * * * * Moreover, it offers a maximum resistance to fire.

"A single instance of failure of any building material is of little interest, owing to the exceptional conditions that may have surrounded it. That is not the situation here. There was not only a general failure to withstand fire, but at least two cases were pitted against one another in the same building, with the concrete doing the work expected of it, and the tile failing."

The great value of concrete as a fireproofing material arises, not from the fact that it best resists heat penetration and surface disintegration, for it may be inferior to other materials in this respect, but that it does not crack seriously, and more often not at all, under the influence of moderately high temperatures, even when suddenly chilled with water; nor does it suffer material expansion. Burnt-clay tile—the chief rival of concrete as a fireproof material—on the other hand, with a co-efficient of expansion twice as great as that of steel and concrete, while it does not disintegrate so readily as concrete under the action of flame, is soon rent asunder by the force of expansion; it cracks and splits, dropping down in great chunks and exposing the parts it is intended to protect to the mercy of the flame, which, if of steel, presently soften, curl up and collapse. That such is the result there is now entirely too much evidence to effectively gainsay, although the advocates of burnt-clay products still stoutly affirm the opposite.

The most nearly fireproof construction results when a fairly rich and fluid mixture is employed, and when by means of smooth, non-absorbent molds and proper working it is given a smooth, hard and impervious surface. Cement-tiles, cast wet in metal molds, and brought to a quick set by introduction of steam into the molds, and given an exceedingly smooth surface by being forced constantly against the metallic surface of the mold as withdrawn (process invented by A. A. Paully, Youngstown, O.), exhibit an extraordinary resistance to fire, enduring red-hot heating and sudden cooling with stream of water without cracking or disintegration. This points a moral: may it not be the porosity— due to air-holes and premature absorption of water from the surface by absorbent molds—of the concrete surface exposed to the fire that facilitates the surface disintegration, allowing the heat to reach all sides of the surface particles simultaneously and thus multiply its effect? This notion is further substantiated by the fact that it is members of the concrete structure that are most exposed to the action of the fire that decompose the fastest: Thus, in a building, the orders in which different parts suffer is: (1) Columns, exposed on all four sides; (2) beams and girders, exposed on bottom and two sides; (3) slabs, exposed generally on the bottom only. In consequence we find this concurrence between the underwriters and the engineering societies— that for columns allowance of not less than 2 in. all around, in beams and girders not less than 1½ in., and in slabs not less than 1 in., should be allowed for fireproofing, the same being not included in the computations as available for carrying load, and no steel being placed closer to the surface than these respective amounts. It is not, of course, feasible to apply Mr. Paully's method to ordinary construction, but it should enforce the necessity of having as smooth and even molds as possible, and if of wood to have same planed and all pores sealed with oil. Thorough wetting, too, of the molds immediately before placing the concrete, is of marked benefit. It is not at all improbable that a method will be developed for applying a glaze to the surface of concrete which will serve the two-fold purpose of enhancing the fire-resistance and the imperviousness. Such a thing seems reasonable.

Permanence:—*That concrete is not durable*—that the material will fatigue, the adhesion existing at the outset between the cement and the steel will eventually cease, and that ultimately disintegration will ensue and probably sudden and disastrous failure. To support this objection there is not at the present time a single instance in the annals of engineering to the author's knowledge. To gainsay it, on the other hand, there are manifold instances—"par example." Of the structures of antiquity those alone of concrete, or the parts of concrete, have survived the ages and are today in good state of preservation. But these are of plain concrete. As to reinforced concrete, structures over a quarter of a century old are still apparently as good as when erected. This is true—that failure for the most part has been confined to the constructional stages, any very serious faults manifesting themselves while yet the structure is in the hands of the builder. And this is indeed fortunate. One may rest assured if the materials are of proper quality and the work of mixing and placing is properly executed, all according to a proper and practical design, there will never be any trouble in this regard.

In regard to the ability of reinforced-concrete to stand prolonged, severe usage, experiments made by and under the direction of Prof. Edgar Marburg, University of Pennsylvania, and reported to the American Society for Testing Materials in 1908 and 1909, are pertinent. Prof. Marburg subjected a number of plain and reinforced concrete specimens to a great number of repeated loadings, approaching in some cases almost to the limit of strength, and yet at the end, tests to failure showed that the specimens had not suffered appreciably, evincing no fatigue. Quoting from the experimentor's conclusions: "The tests made this year substantiate the conclusions indicated in last year's report. They show that one and three quarter million applications of loads causing high working stresses in a reinforced-concrete beam do not materially affect the ultimate strength, the maximum deflection, the bond, the position of the neutral axis, nor the amount of ultimate deformation in the concrete."

Costliness:—*That concrete is not economical.*—to which the most simple and conclusive reply is that were it not economical its use would not be extending at so marvelously rapid a rate. It has been shown why reinforced-concrete is cheaper that either one material alone. The chief reason why concrete itself is cheap is because the materials out of which it is made are among the commonest and therefore the cheapest constituents of the earth's crust. The great item of expense is the centers or molds, the cost of which may exceed 50 per cent, and is seldom less than 30 per cent of the total. In the early days there was great waste in this respect, but now as the result of much careful study on the part of constructors this item has been materially reduced, and it is likely that eventually some type of adjustable, permanent mold—at least more permanent than wood—will be developed that will make for still greater economies. The question of molds is the one item that may prevent concrete from standing in on the basis of first cost. On basis of final cost—which is the only real basis for comparison—weighing in the balance low insurance, minimum maintenance, and minimum depreciation,—not to mention the beneficial effects on the health of the occupants, and on the operating costs, which come of a structure perfectly sweet, clean and sanitary, sound, fire, water, dust, damp and vermin-proof,—warm in winter and cool in summer,—concrete is beyond competition for a substantial and permanent construction.

Viewed from the standpoint of economy the most important factor of concrete construction is the centering or molds. It is so important that the question of an economical centering nearly always has precedence over that of the permanent structure, and in all cases a careful consideration of the one with reference to the other is necessary. A little ingenuity herewith on the part of the designer may—without sacrificing strength—effect a saving that will sway the balance in close competition with steel and timber in favor of concrete. This question should be taken up in detail separately. There is wide latitude for development in the matter of centering concrete.

THE PRINCIPAL PROPERTIES OF THE INGREDIENTS OF CONCRETE:

Foreword:—*In the articles following, a discussion of the materials for structural-concrete, with regard to their principal properties, will be given. The viewpoint will be that of the user who desires to pass an intelligent judgment upon the quality, therefore the fitness, of the materials at the basis of his art. In order to make such inspection and tests as must necessarily preclude the acceptance or rejection of materials, an intimate practical knowledge of their determining properties is manifestly per-requisite. The materials will be considered in the following order: (1) Cement; (2) fine aggregates; (3) coarse aggregates; (4) water; (5) reinforcing-steel.

Classification of Properties:—According to the commonly accepted standards the quality of cement is considered under the heads *Fineness; Setting Qualities; Heaviness,* or *Specific Gravity; Soundness; Tensile Strength; Crushing* or *Compressive Strength; Purity; Chemical Composition.* Some of these include or check others, and not all are subject to exact determination. Oddly enough the two relatively least important, —viz., the fineness and specific gravity, admit of most precise determination, while the *most important one*—the *soundness,* is most difficult to determine. In fact this seems to be true—the order of importance is in exact inverse to the order of ease and preciseness of determination. Arranged in order of ease of determination: (1) Specific Gravity; (2) Fineness; (3) Time of Set; (4) Tensile and Crushing Strength; (5) Soundness. The other two properties enumerated above,—namely, the Purity and Chemical Composition, are really governed or disclosed by the other considerations, and are seldom reported in tests. The properties will be taken up in the order as first given.

The Specific Gravity:—Relatively little importance is now attached to the determination of the specific gravity of cements, although it admits of the most precise determination. It is a quality that varies with age, inasmuch as it depends on

---

*[Compare here the detailed Laboratory work of Chapter VI.]

the amount of moisture that may be present by absorption in the cement, hence the older the cement—by reason of longer exposure to the atmosphere—the greater the absorption of water and therefore the less the specific gravity, since specific gravity is the measure of heaviness as compared with water. Normally, Portland cement is a little more than three times as heavy as water,—that is, an equal volume weighs three times as much,—that is, the specific gravity is three and a fraction. If it were necessary to determine whether a cement was a genuine Portland, this determination would be of avail, inasmuch as natural cements are lighter than Portlands, their specific gravity ranging around 2.75 (see Chapter II for differentiation of Portland cement from other cements). A light specific gravity would also reveal underburning in the kiln,—that is, an incompletely combined cement,—which is, of course, an unsound cement; this being covered sufficiently by the determination for *soundness* the S. G. determination is superfluous; moreover, there would need to be a considerable amount of underburnt particles to affect appreciably this determination. A light specific gravity may also indicate adulteration, but the tests for *soundness* and *strength* reveal this more surely, and for actual demonstration of adulteration the *acid* test (to be described) is most positive. Considered in connection with other tests, or the indications traced out by other means, the *specific gravity* determination becomes an important one; of and by itself, however, it is almost valueless.

The Fineness—Requirements and Value Thereof:—The fineness of cement is readily determined by sifting a quantity through a series of sieves of varying size mesh; it is a purely relative determination and is based upon arbitrary standards established by practice. Generally speaking the greater the fineness the better the cement as cement; that is, the cement the largest proportion of which passed the finest sieve will rank the highest. There is a practical limit to the degree of fineness established by the processes of manufacture, as was recounted in the chapter on Manufacture. Low rank in fineness, while it signifies nothing as to the purity of the cement, or the thoroughness of burning, will yet debar it from high-grade work. Coarseness may be caused by: (1) Overburning, which lowers the efficiency of the grinders, completely vitrified particles being very difficult to reduce; (2) insufficient time in fine grinding mills for material to be all fully reduced; and (3) inefficient or out-of-date grinders in the finishing end.

A certain degree of fineness is requisite because apparently only the extremely fine particles of cement possess cementitious value, coarser particles constituting so much inert matter which acts as filler only, reducing the sand-carrying capacity of the product.

It is probable that these coarse particles do hydrate or mix with water, but very slowly, so that in the ordinary run of events they seem inert. As if to attest this, coarsely ground cements although at first showing lower results, often ultimately show the greatest strength. They may, on the other hand, embody unsound elements which will ultimately, on hydration, cause expansion and possible disintegration. If so, it will be revealed by the accelerated tests for soundness to be hereafter described.

A likely explanation of the inferiority, in fact danger of coarsely ground cement is offered by Dr. W. Michaelis* (see paragraphs on *Set*), which is that the hardening takes place by a process of absorption, part of the cement grains forming at first a gelatinous substance enclosing a heart of un-hydrated matter and the hardening of the gel resulting from absorption by the heart of moisture from it, the gel being thereby gradually converted into a solid. Now, if the grains are a proper fineness the absorption of moisture into the fresh heart, changes it gradually into a gel itself; and thus increasing its volume, such swelling of the interior will act only to increase the tightness of the fit of heart and envelope; but if too coarse the swelling may be sufficient to burst the envelope and thus produce the phenomenon of swelling and blowing characteristic of inferiorly ground cements.

The effect of an excess of coarse particles will be indicated by the tensile strength determinations, since, constituting, in the beginning at least, so much fine sand or stone dust, they will enhance the showing of the neat cement tests but lower the sand-cement. Inasmuch as it is the sand-carrying capacity that counts, such a showing will be adverse to the quality of the cement.

Practice has established that a cement, to rank first-quality, shall pass at least 92 per cent through a No. 100 sieve, and at least 75 per cent through a No. 200. These are low limits and are usually considerably exceeded by standard brands. The finest-ground cement on the market at the present time shows:

Passing No. 100 sieve......................99 plus per cent.
Passing No. 200 sieve......................91 plus per cent.

As recommended by the American Society of Civil Engineers, these sieves should be made of cloth of brass wire. of diameter:
No. 100....................0.0045 in., that is about 1/222 in.
No. 200....................0.0024 in., that is about 1/417 in.
and that the spacing of the mesh should be regular and uniform, and within the following limits:

No. 100, 96 to 100 meshes to the linear inch, that is, about 10,000 to the square inch of surface; No. 200, 188 to 200 meshes to the linear inch, that is, about 40,000 meshes to the square inch.

Wherewith Concerned:—It has been seen that the *fineness* concerns: (1) the cementing value; (2) the density; (3) strength; (4) adhesion of concrete to steel; (5) protection of steel; (6) fluidity of mixture; and therefore the efficiency and economy of the concrete. For proper conditions to obtain,—that is, for

---

*See also Appendix A.

the cement to best fulfill its *essential function* as a *binder* and best mix with the water, producing a mixture of most perfect fluidity,—the cement should be as fine as possible.

THE SETTING QUALITIES OF CEMENT.

Definition:—Cement when mixed with, or rather in, water, almost immediately enters upon a period of chemical change, during which there is a steady growth of crystals or formation of new and insoluble compounds. This growth, once initiated, continues progressively, but at a diminishing rate, throughout a number of years. By it the mass, and whatever is intermingled with it, is gradually converted into an extremely hard solid. The growth begins to manifest itself very shortly upon the addition of water, provided other conditions are favorable, and in the course of a few hours the mass is hard and stiff enough to bear some slight pressure without distortion. The determination of *"setting qualities"* is to ascertain the probable behavior of the cement proposed in any case for use under the average conditions; how soon the stiffening begins, at what rate it proceeds, etc.

The process of setting up probably proceeds something as follows: The cement with water forms a gelatinous substance occupying an increased volume; soon this jelly, by the formation of crystals within itself and the loss of water, begins to harden and as it hardens contracts; the result is the drawing together of all particles enveloped in the gel and their steady conversion into the solidified state, the action being analogous to that of ordinary glue or mucilage; the hardening process continues indefinitely, the eventual being that all calcium present is transformed into carbonate (limestone) and the silica into hydrates (calcium and silica being the two essential elements of Portland cement),—which would explain the progressive and indefinite increase in strength of cement mixtures.   „

The measure of the setting qualities is generally, although not invariably, the measure of the rate of hardening and the gain in strength. There is, in this respect, a marked difference in the behavior of natural cements and Portland cements. The former are found usually to harden very slowly, remaining soft and cheesy often for several days, although they may show on testing a quick set; the latter, on the other hand, may harden up very promptly and yet show a very slow set; hence, by comparison with natural cement, Portland may be characterized as a "prompt hardening and slow setting cement." What causes this difference is not germane, since it does not especially interfere with the ultimate results, but it is worth while noting in case one is confronted with the opposite phenomenon at any time. It is one way of distinguishing between the two cements.

Not all cements, by any means, harden or set up at the

same rate; there are quick-setting cements and there are slow-setting cements. Slow-setting cements are usually better than those that set more rapidly, and a very quick-setting variety would be practically worthless. The trouble with too quick-setting cements, which are also cements taking a high early strength, is that they are prone to markedly decline in strength in the course of two or three months, and may or may not again recover. This behavior is usually, but not always, an indication that the lime or magnesia content, or both, is too high, hence that the cement is unbalanced, or unsound. The strength tests furnish of course the most positive determination of this quality, for one may by no means be sure a cement is inferior because it exhibits a quick set.

Means and Methods of Regulation:—*Cement*, when freshly ground, may exhibit an immediate set and yet be otherwise a perfectly good cement; still, setting so quickly, it would be worthless practically. To regulate the set manufacturers make a practice of adding a slight amount of some foreign material, usually ground gypsum (calcium sulphate), which acts as a retarder. It is added either to the clinker as the latter is fed into the grinders or mingled with the ground cement immediately prior to packing for shipment; the former method is preferable insomuch as it is thus far more likely to be intimately intermingled than if added promiscuously afterwards. The limiting amount of gypsum is approximately 2 per cent; in excess of that amount its effect begins to be injurious, lowering the strength and introducing an expansive agency into the compound. The following table shows the effects of adding varying percentages of this substance:*

Ground gypsum to total cement.	Cement to sand.	Ratio of strength in lbs. per sq. in. for age:		
		7 days.	6 mos.	1 year.
0 per cent	1 : 0	487	743	...
1 per cent	1 : 0	626	746	...
2 per cent	1 : 0	600	754	...
3 per cent	1 : 0	519	742	...
6 per cent	1 : 0	380	660	...
0 per cent	1 : 2	323	492	487
1 per cent	1 : 2	388	530	515
2 per cent	1 : 2	360	547	610
3 per cent	1 : 2	289	607	588
6 per cent	1 : 2	192	663	647

The chief ways in which an attempt is made to retard the set in use at the present time, in addition to the use of gypsum, are: (1) Steaming the cement during the operation of final fine grinding, the action being to reduce the ultra-active properties of the fresh cement to which quick-setting is mostly due; (2) watering the clinker as it emerges from the kiln, for the same reason as afore; (3) aerating the ground cement, again for same purpose. But not one of these seems alone sufficient to the purpose, requiring also the addition of gypsum. By some combi-

*Portland cememt; natural Point aux Pins sand; each result average of 5 specimens. Taken from report of Chief of Engineers, U. S. Army, 1896, p. 2832. Note also remarks on this in Chapters III, IV and V.

nation of these means almost any desired degree of retardation of the set is possible. That combination is best which reduces the amount of gypsum to a minimum.

The action of the gypsum is supposed to be purely mechanical, and probably occurs something as follows: The powdered gypsum when mixed with water tends to coagulate the cement, forming a sort of coating around the minute particles, which delays the attack of the water and so retards the hydration. As if by way of verification, plaster of Paris, which is gypsum devoid of its water of crystallization, is found to be much more efficient than the plain gypsum as a retarder of the set, about half the amount doing the same work. Plaster of Paris, consequently, is being much substituted.

This material, in large percentages, is quite detrimental, tending to make the cement blow and swell much like it does under the action of sea water, and for a similar reason—that it probably forms with the aluminous element of the cement an expansive compound, tending to congest the cement in the mixture and thus interfere with its perfect liquification, it is in all events an evil, and the necessity for using it to be regretted. It is not at all improbable that a better and safer way of doing the same thing will ultimately be developed.

Another explanation of the probable action of gypsum in delaying the set of Portland cement is offered by Dr. W. Michaelis (a German chemist of no little repute in the cement industry). His explanation is, that the addition of certain compounds like gypsum and calcium chloride, themselves strongly crystalline in their tendency, to a solution tending naturally more to the colloidal state (jelly or glue-like), tends to counteract the formation of colloids—which takes place some little time before crystals begin to form—and so to retard the hardening. So, contrarywise, by addition of compounds of the opposite nature, that is tending themselves strongly to the colloidal condition as they solidify, as for instance sodium silicate or water glass, an acceleration of the set is produced.

It should be mentioned in this connection that, according to Dr. Michaelis, the phenomenon of setting and hardening of Portland cement mixtures is due to the formation of colloids rather than crystals,—in other words that Portland cement mixed with water acts like a mineral glue, drawing all particles together as it congeals or stiffens, and finally becoming very hard, imparts solidity to the entire mass. In this opinion Dr. Michaelis has many concurrers. For a complete discussion of the same the student is referred to CONCRETE ENGINEERING for November, 1909.*

Other Modifiers of Set:—*The rate of set* is influenced by the chemical composition and the degree of burning; by the fineness of grinding; by the manipulation of the product after

*This has been included in this volume as Appendix A.

grinding; and by physical conditions surrounding its use. For the influence of the chemical composition the reader is referred back to the section on chemical composition. In general under-burnt cements, as well as freshly-burnt cements, set quicker; but not invariably so. High-limed and high-alumina cements also tend to be more quick-setting. The finer the cement the quicker the rate of set, since the hydration is less delayed; it is the coarse particles that hydrate most slowly. The amount of aeration given the freshly ground product markedly influences the rate of setting, the more complete the more it is improved.

Physical conditions that exert a marked influence upon the *rate of set* are: (1) The temperature; (2) the degree of wetness or amount of water used in mixing; (3) the quantity of material mixed at a time; (4) the quiescence of the mass after mixed.

Temperature:—Cement ceases to set up at the freezing point of water, for the reason that the water at that point, instead of entering into crystalline structure with the cement, forms an independent crystalline structure,—that is, becomes ice. Conversely, the hotter the temperature, up to the boiling point of water, the more the set is accelerated. This has a very important bearing upon practical operations. Consider its effect first in the heat of summer, when, especially around midday, the sun is very warm. Under such conditions the set may be accelerated so that the mixture will become stiff while yet being mixed; or at any rate before it can be poured into the molds and tamped. It then becomes necessary to adopt some measures to retard the set,—or, which is the same in effect, to guard the fresh mixture from the boiling rays. This may be accomplished by making the mixture rather wetter than necessary, covering the vehicles in which the fresh mixture is being transported with wet cloths, and either covering the concrete in place similarly or protecting the entire scene of operations with awnings. These means will be given some further consideration in other works.

Freezing Temperatures:—In case the temperature is hovering around freezing point, it is necessary to do something to encourage the setting; practically the only way to do this is to apply heat. Heat may be applied in several ways, or combination of several: (1) Heating the raw materials before mixing and using hot water in mix;* (2) insulating the freshly deposited concrete from effect of exterior temperature conditions by covering with a suitable material, when—unless the temperature be considerably below 32 degrees F.,—the heat generated by the chemical action beginning to take place with the mass,— that is, the heat given out by the crystallization of the cement. will be sufficient to produce a normal rate of setting; (3) insulating the top surface and applying artificial heat to the under sur-

*Note in this connection the method of applying steam into drum of mixer while mixing. See Concrete Engineering, December, 1909.

Salt may be added to the water in mixing, but this in no way alters the set, simply lowering the freezing point of the water and thereby making it possible to carry on operations at a lower temperature, enabling one possibly to dispense with heating of the raw materials and affording more leeway in manipulating. Calcium chloride also lowers the freezing point and is by some preferred to salt on the ground of its greater affinity for cement; but if less alien than salt its tendency on the contrary, to retard the set as above explained, and thus to prolong the danger of freezing, would act adversely. Below 22 degrees F., even using salt, it is necessary to heat materials, apply protection above and heat below the freshly deposited concrete, if a normal setting up of the cement is desired. Concrete is often . mixed in cold weather without any safeguard whatsoever, packed into the molds and allowed to freeze solid; then, when the weather moderates, the frozen stuff softens up and the set goes on as if nothing unusual had intervened. This method is questionable, for if another freeze ensues before a hard set has been taken, the alternate freeze and thaw very effectually ruins green concrete. It is likewise bad, unless one has plenty of time to wait for the structure until thawing out shall have taken place, and has the money to tie up in the falsework in the meantime, as obviously the centers cannot be removed while the concrete is in a frozen condition. There is one vital objection, in any event, to frozen concrete, namely that, as water in turning to ice expands, hence the structure of the mass is bound to be distended during the freezing process, which means when the water has again assumed its normal condition there will be voids left, representing the difference in volume between water frozen and water liquid. To a certain extent, upon the thawing of frozen concrete, it will of its own gravity tend to correct such voids, but by no means positively, hence frozen concrete will always be more porous than normally, which, in the case of a surface exposed to wear, or acting to protect embedded metal, or to resist water pressure, is a positive detriment and to be avoided.

It is thus seen that concreting under either extreme of temperature is fraught with difficulties and dangers that call for special attention.

Hardening at Low Temperatures:—At temperatures between 32 and 40 degrees F., the hardening of cement progresses but slowly, and owing to the retarded formation of colloidal and crystalline compounds, to which the hardening is due, certain peculiar compounds of a pulverulent nature are sometimes formed. This action is more likely to take place if the atmosphere is surcharged with carbonic acid. The result is at any rate that the surface, for the depth of a quarter-inch or so subsequently shows friability. Below this, normal conditions would obtain as usual. Such an action would be especially liable to take place if, as is often the circumstance in cold weather, coke or coal furnaces were used to warm the atmosphere of an enclosure where fresh concrete had lately been deposited, since the burning of fuel makes carbonic-acid gas (CO_2) very rapidly, and being heavier than air it sinks to the level of the floor, thereby coming into close contact with the raw cement surface. Therefore, such devices should not be used for heating unless it is certain that a temperature of at least 45 degrees can be maintained during the critical stage of setting.

A similar action might occur in event of concreting being carried on at low temperatures in manufacturing districts, where the air is heavy-laden with smoke and carbon dioxide: in the open, fresh atmosphere it would be less likely to occur—if at all. This is a condition, evidently, to be borne in mind. While not seriously affecting the strength of the concrete it may cause excessive dusting, or crumbling even, of the surface, hence is to be avoided.

M. Le Chatelier, in his classic "Constitution of Hydraulic Mortars," remarked this phenomenon as follows:

"* * A hydrate of lime, $CaCO_3.5H_2O$ exists which can be produced by the carbonation of lime below 41 degrees F., but it is very unstable and the least rise in temperature transforms it into a pulverulent anhydrous calcium carbonate, devoid, consequently, of all strength. It may be useful to take account of this fact in the hardening of cements at low temperatures, especially when exposed to the air. * * * * *"

Most Favorable Temperature:—*The most favorable temperature* for the setting of cement is around 70 degrees F. Then no especial precautions either way are required. In the northern latitudes of the United States, and in the southern part of Canada, the golden months for concreting out-of-doors are April to October, and occasionally right through till the last of December, although during the months of November and December it may be and usually is necessary to protect the concrete from frost the first night after deposital; also during July and August, especially around noon-time, the fresh concrete must be guarded from the direct rays of the sun, and too rapid evaporation of the water prevented by swathing with wet cloths, or moist sand or sawdust, and after hardened keeping well wetted for several days,—the longer the better for the strength of the concrete. During the early spring months, when the temperature is mild and the atmosphere moist, no special precautions are necessary, although watering of the new surface is to be observed as usual. The determination of setting qualities is made usually at about 70 degrees F.

Relation of Setting Qualities to Time of Year:—Obviously, a quicker setting cement may be used, and generally should be, in cold weather than in warm; conversely a cement for summer use should be as slow setting as possible. Most specifications make a differentiation of this kind.

Effect of Mechanical Agitation and Regaging:—One of the ways mentioned for retarding the set was agitating the mass after mixed,—that is, not allowing it to be quiescent. The set evidently cannot begin while the mixture is constantly in a state of motion. The harm is not so great if the mass is not allowed to stiffen during an interval of quiet; if stiffening has begun and the mixture is regaged by adding more water and agitating again the strength suffers. It is strongly advisable not to disturb the mixture after material stiffening has begun, as the strength is then bound to be reduced. In the "Report of Watertown Arsenal Tests," for 1901, results of experiments are presented which show these facts clearly. Witness: A sample of cement, placed in the mold directly after mixing, exhibited a compressive strength of 3,549 lb. per sq. in.; a sample of the same after being regaged once every hour for 20 successive hours, fell to 1,690, a loss in strength of over 50 per cent. Another example in the same report, under the condition that the mixture was kept in a continual state of agitation, the sample taken out from the fresh mixture tested up to 5,457 lb.; that at the end of one hour of agitation, 4,665; two hours, 6,421; four hours, 5,470; six hours, 2,718; 20 hours, 1,901; 50 hours 1,168; 100 hours, 681. The greatest drop, it is seen, happened after the fourth hour, and was very rapid thereafter. These results are not, of course, of great practical importance, as it is very unusual that even as much as an hour delay ensues between the initial mixing and the final placing. But it does show that no material should be left unused over night and then regaged in the morning and used; such practice is pernicious and must not be permitted.

Avoidance of Shocks:—It is also evident that the mixture, after once poured into the molds, should be immediately agitated all that may be necessary to settle it into position, release air bubbles and secure a dense mass and smooth face. Then it *should* be left entirely alone for at least a week, if possible, and the surface adequately guarded from the transit of the working men; also from shocks of all manner, either imparted by direct contact above or by disturbal of the props beneath. Keep off and keep away! The only workman allowed on the surface is he whose business is to keep it wet, and he should be careful to walk on planks.

Periods of Set:—In testing the period of set is divided into two purely conventional parts: (1) The *initial* set and (2) the *final set*. As previously mentioned, the cement, once mixed with water, immediately, under normal conditions, begins to congeal or set up. How fast this stiffening proceeds we have seen depends on the temperature, the amount or proportion of water used in mixing, the body of cement mixed, and how long agitated.

For the purpose of testing certain of these conditions are arbitrarily fixed: The temperature at 68 to 70 degrees F.; the percentage of water at about 25 (subject to trial); the period of *initial set*,* or when material first evinces appreciable evidence of hardening, as the point when a 1/12-in. wire weighted with a ¼-lb. cease to penetrate; and the *final set*, or when the material has become solid enough to bear slight pressure without distortion, as the point when a 1/24-in. wire or needle weighted with 1 lb. cease to penetrate. The one variable condition which it is not possible to fix is the personality of the tester. The carefulness and thoroughness with which all conditions are observed, the thoroughness of the mixing, and the accuracy of observation,—all factors influencing results, are herewith concerned. No two testers can ever be reasonably expected to agree on the results. A rough way of determining the periods of set, in case the needles are lacking (they are known as the Gilmore needles) is with the finger nail; when it cuts in easily and yet leaves a clear impress the initial set may be said to have been reached; when the nail cease to indent the surface, the final set. It is usually specified that Portland cements shall take

> Initial set in from 30 minutes to two hours.
> Final set in from one hour to eight hours.

The amount of water is determined by the resulting consistency of the mass, what is called the *normal-consistency*† being reached when a condition of plasticity ensues; then a ball of it in the hand is like sticky blue clay, and may be tossed from one hand to the other or dropped several feet without separating. Usually about 25 per cent by weight of water is required to produce this consistency. If stiffer than this, the set will show quicker; if wetter, more delayed. There need be no alarm generally if concrete deposited near the end of a day, especially in cool, damp weather, is still soft the next morning; it will harden up as the day warms up. Then again the coming of a heavy, prolonged rain may, by par-wetting the freshly-deposited mass, keep it in a soft, cheesy condition for a couple of days, but it will finally set up and be none the worse for the delay, unless indeed the rain has been severe enough to wash some of it away, or the forms so leaky as to permit of the cement leaching out; a light rain is usually harmless, rather a benefit; but a heavy, pounding rain may materially injure the surface,—from such it is wise to protect the green concrete.

---

*The method of determining the periods of initial and final set here indicated is that invented by General Gilmore, of the United States Army Engineers. The method used abroad and being adopted in the laboratories here. in accordance with the recommendations of the American Society of Civil Engineers, is that known as the Vicat Needle Apparatus, a description of which will be given in another place. In effect they are practically the same.

†The Normal Consistency Standard recommended by the American Society of Civil Engineers is thus described: "The cement paste is of Normal Consistency when the cylinder (of the Vicat Needle Apparatus) penetrates to a point in the mass 10 mm. (0.39″) below the top of the ring. Great care must be taken to fill the ring exactly to the top." Description to be given subsequently.
See Chapter VI, under this heading.

Integretation of Indications of Tests:—The rate of setting, as determined by test, may afford little clue as to the behavior of the cement as concrete under perhaps very different conditions; but such results coupled with proper exercise of judgment, as comes only by experience, lead to valuable inferences.

Nor is the rate of set an accurate gage of the rate of gain of strength, and hence useful in deciding how soon new work may be uncentered and put into use; for slow setting cements may make the most substantial gain during the critical period, and conversely quick setting cements make the slowest gain. Usually cements that harden slowly reach the greatest ultimate strength.

Precautions to Procure Uniformity of Setting:—As already mentioned, the setting once begun continues progressively for an indefienite period—indeed. probably for many years; but it proceeds most rapidly during the first few days. This, then, is the critical period, and during it the mass must not only be undisturbed, but it must be kept constantly moist—otherwise setting will not progress favorably and the surface especially will suffer for lack of sufficient water. There are several ways of providing for this need; perhaps the most satisfactory is to cover the fresh surface, as soon as it is hard enough to bear it, with a layer of wet sand, say about an inch thick. This retains moisture excellently, thereby nicely wet-blanketing the concrete and feeding it with water as it so requires. Failure to observe some such precaution is almost certain to result in a weakening of the concrete. due to the setting up of distortionate hardening stresses, and especially in hot, dry weather is the danger real, the outer crust being then often times seriously damaged. For proper and safe results it is absolutely essential that uniform conditions be maintained for at least five days, and the longer thereafter the better. Attesting the benefit of wet sand, concrete cubes for testing purposes, molded in sand and kept moistened until tested at the end of a month, showed percentages of increase in crushing strength as high as 50 per çent.

Effect of Amount of Water Used in Gaging:—The following table shows effect of varying percentages of water on the set: (Taken from Watertown Arsenal Tests, 1901.)

Name of brand.	Water. Per cent.	Gillmore method.			
		Initial set. Hrs.	Min.	Final set. Hrs.	Min.
Alpha Portland Cement	20	2	20	5	0
Alpha Portland Cement	25	3	20	7	30
Alpha Portland Cement	30	5	40	..	..
Atlas Portland Cement	20	4	05	7	10
Atlas Portland Cement	25	5	10	8	05
Atlas Portland Cement	30	7	00	..	..

Chemical Heat of Setting:—The following table, also abstracted from the same reports, shows the temperature acquired by Portland cements in setting:

Name of brand.	Water. Per cent.	Max. temp., degrees Cent.	Ultimate† crushing res.	Age in days.
Alpha Portland	26.2	95	5,706*	9
Star Portland	26.5	76	.....	..
Storm King Portland	27.0	42.5	.....	..
Whitehall Portland	25.2	103.5	.....	..
Dyckerhoff Portland	25.0	63	1,547	13
Atlas Portland	22.7	81.5	4,872	9

Experiments were made with 12-inch cubes, the upper surfaces exposed to the air and the temperature taken by inserting a thermometer bulb in the center of the mass. It is interesting to note that the temperatures attained are very high, in one case exceeding the boiling point of water (water boils at 100 degrees C.). Several hours were consumed in reaching the maximum temperature. Of course this effect would be most pronounced for neat cement and least pronounced for actual concrete; but it is sufficient to pretty well keep the interior of freshly deposited concrete from freezing if the set has once begun in good shape.

TENSILE STRENGTH.

Importance:—The *tensile strength* of Portland cement is, as afore stated, next to the *soundness* or *constancy of volume*, the most important property. It is important because the very integrity and worth of whatever combination the cement may enter into is dependent on the *strength* of its *life-element*—the *cement*—that is, the cement's degree of cohesiveness or cementing value. It is also important because it virtually furnishes an index of every other quality and becomes a main determination itself and a check on all minor determinations of quality; any deficiency whatever in the quality of the material is very sure to show in the tensile strength. This primal quality is affected by, and is therefore a measure of, the chemical composition, the completeness of burning or calcination or combination of the raw materials, the fineness of grinding, the rate of set, and the soundness.

Relation of Tensile to Crushing Strength:—It may be inquired, "Why is the tensile strength important when it is its compressive strength with which we are mainly concerned,— cement is not used in tension?" Because the tensile strength and the compressive strength are in direct proportion, and the former is the more reliable determination, besides ordinarily more easily made. The compressive resistance is usually from 6 to 10 times the tensile strength; hence if the latter be known the former can be readily calculated with sufficient accuracy. (See table at end of this article.) The tensile determination, moreover, is entirely rational, for in the last analysis a failure in compression is a

---

*Not ruptured. †Lbs. per sq. in.

failure in tension; crushing takes place by the yielding or pulling apart of the particles; it is slower than direct tension because the particles are less free to move, being equally restrained on several sides at once. Formerly the strength was determined in direct compression, and it is to some extent now, but more as a check on the tensile determination, since the latter is the commonly accepted standard. In the laboratory section of this work a convenient and simple method is outlined for testing in direct compression, which has been found by the authors to give at least as satisfactory results and with less trouble and expense; it is advanced with the hope that it may meet with wide-spread adoption. The discussion of the subject here, however, will be from the standpoint of tensile strength.

Rate of Gains of Strength:—Because a cement appears to harden and gain strength very slowly during the early stages is no indication of inferiority; the most reliable cements, on the contrary, will exhibit just this sort of gain, and they will usually be found to gain progressively, without periods of reversal or stagnation, and attain ultimately the greatest strength.

Cements, on the other hand, that attain great initial strength and hardness will usually be found to exhibit a retrogression in from 2 to 3 months, which is just the time when the gain should be steadiest, and although they may recover and go on gaining, they never attain to the ultimate strength of the other. Very often a high early showing is an evidence of "doctoring-up," the manufacturer having done something to his product to boost it at the start in order to cloak an inferiority which would bar it. Hence high-testing cements may at times be regarded with suspicion.

A proper cement will continue to gain steadily for an indefinite period; indeed, tests made for periods extending over 20 years still show gaining. However, a practical maximum is attained in from 60 to 90 days, and it is upon the strength to be expected at this period that calculations of design are based. The practical maximum may not be actually more than 2/3 to 3/4 the real or final ultimate. The rate of gain normally is such that in from 4 to 6 weeks about 3/4 of the practical maximum is reached, at the end of which time, therefore, the structure may safely be put into use; but it should not be fully loaded until after three to six months. Concrete is unique in this respect, for with all other structural materials the initial strength is the greatest and a retrogression almost immediately sets in, requiring replacement eventually and lightening loads as the years multiply; precisely the opposite is true with concrete, a fact which may well be taken advantage of in the calculations.

Test loads on structures are usually specified at the end of 60 or 90 days; evidently the longer postponed the severer may be the exactions prescribed. The laboratory tests for tensile strength are valuable in themselves as indicating the value of the cement; but less valuable in indicating the actual gain of strength of the structure. To keep mark of this it is desirable to mold cubes (say 6 inches on a side) from each day's work and to test these in periods of seven days, two weeks, one month, two months and three months; that is five cubes should be saved out of each day's work, molded to true and accurate alignment, in smooth molds, preferably of metal, and treated to the same conditions of temperature and moisture as the actual work. Then compressive tests of these at the stated intervals will furnish very valuable indications and a safe guide as to the removal of centers.

The following table is of interest as showing the behavior of cement in tensile strength. The period of reversal occurs for nearly all brands between 2 and 6 months, but in nearly all

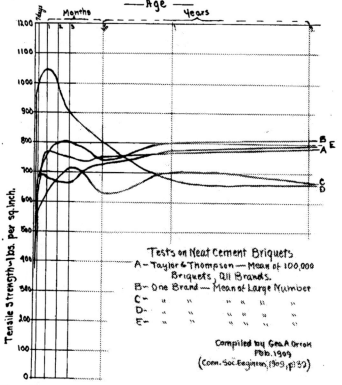

FIG. 42—CURVES SHOWING LONG TIME COMPARATIVE TESTS ON VARIOUS BRANDS OF CEMENT.

cases there is a recovery after 3 to 6 months and the gain is thereafter steady and gradual, and the final strength reached the same, except in the curves C and D. Cements of this character are dangerous and should be rejected. The curve drawn from Taylor and Thompson is especially valuable in so much as it is the mean of a very large number of tests on all kinds of cement, and may therefore be taken as the general behavior of American Portland cements.

Permanence of Strength:—Some long time tests of the tensile strength of cement exhibit a peculiar behavior after a certain age in the shape of a retrograde movement of the tensile strength. According to one of the best authorities, namely the United States Geological Survey reports, the majority of tests they have made exhibit this tendency, the retrogression occurring from 90 to 180 days. Up till that period there seems to be a steady gain; and after that almost constant strength up till a year; then a recovery and a very gradual gain in strength for an indefinite period. Mention of this phenomenon seems called for insomuch as there has been in the past, and there is liable to be in the future, considerable alarm and therefore doubt as to the permanence of all cement structures by the promiscuous dissemination of results of this kind on the part of those not understanding to those also not understanding the full facts. Reference is made, in this connection, to some valuable data published in the May, 1909, issue of CONCRETE ENGINEERING, by Mr. A. E. Smith, of Pittsburg, being a compilation of results obtained by him and by Mr. S. S. Ferguson, M. E., C. E., also of Pittsburg, extending over a period of 20 years; the results are the more valuable insomuch as they are made mostly on old brands which are inferior to those in use today and from work under the author's own supervision. The oldest sample, 20 years, 11 months, being one of eight briquets, seven of which were broken at six months and averaged 500 pounds per square inch tensile strength, showed a strength of 1,126 pounds, a gain of over 120 per cent over the strength at six months. This sample was of one part Alsen Portland cement to one part river sand. Other of the same lot of tests show proportionate results, all combining to prove that with an increase in age there is a corresponding increase in the strength; and leaving little room for doubt as to the permanency of Portland cement.

Tensile Strength Determinations:—*Tensile strength determinations* are of two kinds: (1) those measuring the strength of the neat or clear cement; (2) those measuring the sand-carrying capacity. Both are valuable, and if there is any choice as to relative importance the sand determination should have the preference—for the reason that it more nearly conforms to the conditions of practice. It is customary to make both determinations and to arrive at a fair judgment by mutual comparison. This is true—that the cement exhibiting the greatest strength

neat is not necessarily the best cement for concrete, for in combination with sand it may exhibit the poorest results—and it is
the latter that in any case should be accepted as the final criterion.
The explanation in this case is simple. The superior showing of
the neat cement is due to the presence of coarse, inert particles
which really act as so much sand, producing a better balanced
blend and therefore testing up higher; whereas, mixed with sand,
the coarse particles acting now to reduce the sand carrying capacity of the cement, a contrary result is obtained. Thus is inferior degree of fineness or adulteration with pulverized rock revealed.

Significance of Determinations:—Failure to satisfy
either the neat cement or the sand and cement standard usually
signifies *unsoundness*—that is, improper chemical composition, or
incomplete amalgamation, since particles either wrongly combined
or not thoroughly combined have little or no cementitious value.
It may also signify *adulteration;* a question which will be discussed separately. In any case the precise difficulty may require
the digestion of the facts of several determinations and then even
not be clear. Again, the results may go by inversion. W. Purvis
Taylor, in charge of the Testing Laboratories of the city of
Philadelphia, relates the case of a curious discrepancy; "Our
high record cube, breaking at over 5,300 pounds per square inch,
was made from a cement that barely passed the tensile strength
requirements of the specifications, and about which there was
some question whether it would be allowed to be used; and there
are many discrepancies of this sort." Which goes to show that
the determination of tensile strength, although one of the decisive
determinations upon which the acceptance or rejection of a cement
is based, is not altogether reliable.

Personal Equation:—Here, also, even more than with
the determination of the set, does the personal equation figure,
for it concerns not only the preparation of the test pieces but their
breaking. Different degrees of manipulation of the cement-putty
in the test-piece molds ("briquet-molds") may give widely varying results. Again, in applying the breaking load in the testing
machines, a little unsteadiness on the part of the operator may
easily cause the piece to fracture at greatly reduced load. Even
the same operator under the same conditions cannot be relied
upon to get absolutely consistent results, since "personal equation" is itself a variable. There is only one way to reduce the
liability of this error—that is to introduce mechanical molders
and breakers, that by means of delicately adjusted springs will
uniform conditions to a nicety. The human element in the operation would thereby be reduced to a minimum, and therewith the
personal error of determination. Lack, at the present time, of
such appliances, and this fact of non-uniformity of results make
comparison or data of different testers of little absolute value
and tend also to negate the value of data furnished by the mill,

since a little over-zeal on the part of the mill-tester may easily secure for an inferior cement a ranking out of proportion to its actual standard. The user should bear these facts in mind: He should be wary of "too-good" reports from a tester, and on the other hand of rejecting a cement because of results apparently "too-low" or sailing close-hauled.

The Standards of Tensile Strength, as adopted by the American Society for Testing Materials (1909), and generally observed, are:

Neat or Clear Cement:

Age.	Strength.	
One day, in moist air.................................175	225	lbs. per sq. inch
7 days (1 in moist air, 6 in water) 500	600	" " " "
28 days (1 in moist air, 27 in water) 600	700	" " " "

One Part Cement to Three Parts Standard Sand:

Age.	Strength.	
7 days (1 in moist air, 6 in water) 200	250	lbs. per sq. inch.
28 days (1 in moist air, 27 in water) 275	350	" " " "

With the additional requirement that there shall be no intervening periods of retrogression.

These are, however, low limits, and a cement that failed consistently to exceed the maximum requirements would today hardly be considered a first-quality cement. One of the latest and highest testing brands on the market, shows, in one instance:

Neat, 1 day 325	lbs. per sq. inch.	Average 5 specimens
Neat, 7 days 676	" " " "	" " "
1:3, 7 days 255	" " " "	" " "
1:3, 28 days 331	" " " "	" " "

Results of a Long-Time Test:—The following table, taken from Falk's Cements, Mortars, and Concretes, being plot-

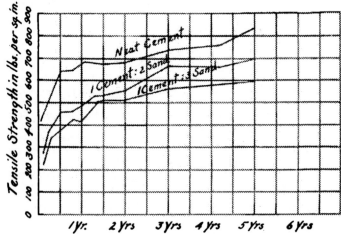

FIG. 43—CURVES SHOWING LONG TIME TESTS ON CEMENT.

ted from the results of long-time tests made on samples of cement representing 300,000 bbls. of the Giant brand of Portland cement, as reported by Robert W. Lesley, in the *Journal of the Association of Engineering Societies,* 1895, represents the behavior of an average good brand of Portland cement. Each point plotted is the average of 1,000 to 1,300 test-pieces. Considering only the 1:3 tests, it is seen that the strength at 3 months, as compared with the strength at 5 years, is 60 per cent, and the strength at 6 months 73 per cent; compared however with the strength at 1 year, the percentages are respectively 82 and 100 per cent. The strength is thus seen to remain at a standstill between 6 months and one year and afterward to increase again indefinitely.

Age of Specimens in Years.

Note: Each point the average of 1,000 to 1,300 specimens.

For 1:3 tests, strength at 3 mons.	60%	strength 5 yrs.
" " " " " 6 "	73%	" 5 "
" " " " " 3 "	82%	" 1 "
" " " " " 6 "	100%	" 1 "

Practical Value of Tests:—It should be remembered that, in all these tests, the figures representing the ultimate strength will only apply relatively to the actual strengths realized, which may run over or under the indication of the test. The more nearly the test conditions approximate actual conditions the more valuable the indication. For this reason the sand-cement tests are a more reliable indication than the neat cement, more nearly approaching the actual mixture. The comparative value is somewhat vitiated by the fact that the *Standard Sand* used in testing is usually quite different from the sand used in practice, containing a larger percentage of voids. The average natural sand, containing usually a small amount of silt and fairly well graded fine to coarse, will give the higher results. The Standard Sand test will, then, over rather than undervalue the quality.

Another factor that will affect the results is the variability in the quantity and temperature of water used. Tests are conducted under uniform conditions—actual work under widely variable conditions, both degree of wetness and temperature varying. Moreover, there is as a rule more water used actually than in test specimens. These will all tend to invalidate deductions based upon early comparisons, although in the long run the strength realized in practice will fulfill fairly well the expectations of tests.

The difference in quantities will also materially affect the results, it being of course possible to obtain much higher values with a small laboratory-made test specimen, perfectly proportioned and thoroughly mixed and tamped, and maintained under ideal conditions, than with a large mass, made under the more uncertain conditions of practice.

It is thus evident that the chief value of test data are for comparison with Standard requirements and other test data, and that as a practical guide they are very relative.

Ratio Between Tensile and Compressive Strength:— The following table, compiled from the Watertown Arsenal Report for 1902, is of interest, showing the ratio under varying conditions of age and wetness between the tensile and compressive strength of neat cement. The brand of cement used was the Peninsula Portland, and each value in the table is the average of 10 determinations. It will be noted that the ratio is never far from 10, at the age of one month, and that the maximum results seem to obtain with 25 per cent water.

Age in		Gauged with 20 per cent. water.			Gauged with 22 per cent. water.			Gauged with 25 per cent. water.		
Air. Days.	Water. Days.	Comp strength, Lbs. per sq. in.	Tens. strength, Lbs. per sq. in.	Ratio.	Comp strength, Lbs. per sq. in.	Tens. strength, Lbs. per sq. in.	Ratio.	Comp strength, Lbs. per sq. in.	Tens. strength, Lbs. per sq. in.	Ratio.
1	..	717	196	3.7	595	189	3.1	430	190	2.3
17	..	3,040	354	8.6	3,260	392	8.3	2,610	402	6.5
28	..	3,990	566	7.1	3,760	457	8.2	3,130	450	7.0
1	6	4,250	780	5.5	4,720	666	5.8	3,880	329	11.8
1	27	7,370	906	8.1	6,870	866	7.9	7,580	758	10.0

Make-Shift Test of Tensile Strength:—Cases may arise where it is impossible to make the tensile tests according to standard specifications, and also impossible to make the equivalent compressive tests. A fair approximation may be attained in the following manner: (Refer to the article giving standard methods of making and molding briquets for testing; the same directions will apply here.)

(a) Build a box of thin planed stuff having a series of partitions dividing it up into compartments 1 inch deep and 1 inch wide and 7 inches long. There will be needed in all about 18 compartments: 12 for neat cement paste and 6 for 1:3 mortar test, which allows 3 test pieces each for the 24 hour 7 day, and 28 days tests for the neat cement and 3 each for the 7 day and 28 day mortar tests. If more test pieces are required than this more compartments will need to be provided. The wood should be well seasoned stock and the walls of the compartments should be well oiled to seal the pores. The dimensions should be as true as it is possible with the means at hand to make them.

(b) Out of sheet steel or hard wood make two knife edges. (Triangular shaped supports.) Same should be approximately 2 inches high with the top edge slightly rounded. Place these firmly on a short piece of 2-inch plank and exactly 6 inches apart, measured from center to center of knife-edges. Place the

board thus equipped on the two supports at the same level and
nail it fast. This will provide a firm, level support for the two
knife edges and allow access underneath. Midway between the
knife edges and exactly on their line of centers bore an inch diameter hole in the board. The apparatus is now ready for testing.

(c) Provide a strong spring balance and a large pail.

(d) Place the test piece, after it has hardened the proper
length of time, upon the knife edges, and center it exactly. It
will overhand the edges a half-inch. At the exact point, 3 inches
either way from a knife edge and directly over the hole in the
board, wrap a stout wire, (about No. 8) carrying the ends hair-
pin like down through the hole and joining them together a
few inches below. Attach the spring balance below, and to the
handle of the spring in turn the pail. The wire where it touches

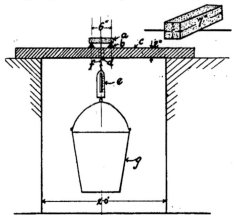

Fig. 44—Field-Made Apparatus for Testing Tensile Strength
of Cement.

the test piece should be prevented from bearing too sharply by
interposing a small strip of wood or metal, preferably metal,
which is just the width of the piece, and the wire below the
test piece should be spread by the insertion of a forked stick,
1 inch between crotches. Great care must be exercised to see
that when everything is in place the balance hangs directly under
the center both ways. Then the load will be applied symmetrically.

(e) The balance will register a small amount at the outset
due to its own weight and the weight of the pail. Read and
record this. Then begin to pour either sand or water (preferably
sand) into the pail, being careful to add it both gently and slowly, and to see that the balance hangs still. When the load has
reached a certain point the test piece will probably fail abruptly
by cracking at or near the center point. If incipient cracks ap-

pear along the under edge before this, note at what loading and record. Record also the final weight, which will be the ultimate or breaking load. The weight of the piece may be neglected.

(f) The tensile strength of the cement can then be calculated as follows: The external force producing bending = ultimate load = P (lb.) This force being concentrated at the mid-point, induces a bending moment (according to mechanics) = ¼ $P \times$ *Span*, the span being the distance between knife edges. Here then $B. M.$ = ¼ $P \times 6$ in. = 3/2 P in. lb., where P is expressed in pounds and a fraction (if balance reads to ounces, reduce to pounds and fraction). This $B. M.$ is resisted by the internal strength of the material, which is known as the Resisting Moment ($R.M.$).

(g). The $R.M.$ of a rectangular section of homogeneous material equals (according to the laws of mechanics) $\dfrac{p \ I'}{d}$, where p = fiber stress in lb. per sq. in. of section, I = the Moment of Inertia in quad inches, and d = the depth of the section in inches. In this instance p is the quantity sought; $I = \dfrac{bd'}{12} = \dfrac{I \times I}{12} =$ 1/12 in.'; and d = 1 in.

Hence $R.M.$ = p $\times$ 2 $\times$ 1/12 $\times$ 1/1 = 1/6 p.

Equating R. M. to $B. M.$, we have,

1/6 p = 3/2 P, or

$$p = 9 \ P \dots\dots\dots\dots\dots\dots\dots\dots\dots\dots\dots\dots\dots\dots (1)$$

The tensile strength of the material is then given directly by inserting in the formula (1) the value in pounds and fraction of the load required to break the test piece.

Suppose, for example, that failure ensued when the balance registered 50 pounds; then inserting this value in (1),

p = 9 $\times$ 50 = 450 pounds.

(h).—Note:—Inasmuch as the strength of the neat cement at 7 days, and certainly at 28 days, may be expected to range between 500 and 800 lb., it will be necessary to provide a spring balance capable of taking at least 100 lb., and to provide also adequate means for loading same. The sand may not be found sufficient, when chunks of iron and lead may be substituted.

(i) Note:—The values hereby obtained, if the work is carefully done, will exceed the values ordinarily obtained in tensile tests, which are naturally somewhat lower because made by direct pull, whereas in this case the piece is in flexure and the pull indirect. Experiments to determine this relation show that the strength by flexure will exceed strength by direct tension from 25 to 50 per cent. In the present instance, considering the crudeness of the apparatus, the lower relation may be assumed.

Hence, if a result of 500 lb. is obtained, it should, for comparison with general data and standard requirements, be multiplied by 1¼, giving a total of 625 lb. .

THE COMPRESSIVE STRENGTH.

Importance: *The compressive strength* is, in a sense. the most important quality of a cement. We have seen, in the previous article, its relation to the tensile strength, and how for convenience the determination of the latter has come to be accepted as sufficient. In the laboratory* section of this volume a simple method of determining the compressive strength directly is outlined. The main reason heretofore for the preference given the tensile determination has been, not its greater desirability, for the reverse is true, but its greater convenience. If as equally convenient a method were available for making the ·compressive determination it would no doubt receive speedy preference. The main facts with regard to the compressive qualities, especially as it bears upon the tensile strength, have already been indicated, but here a few of the more salient features will be given emphasis. .

By What Determined:—The compressive strength of concrete depends mainly but not entirely on that of the cement, or the cement and sand. Given a thorough mix and aggregate of good quality, and the compressive strength will depend mostly on that of the cement; let the aggregate itself be low in compressive strength, lower than that of the cement, and it will be the strength of the aggregate and not that of the cement that will determine. These are points that will be dwelt upon further in the consideration of the qualities of aggregates.

Value of Compressive Tests:—The more nearly the conditions of a test approach the conditions of practice, obviously the greater its value as a positive indication. Tests for compressive strength should, therefore, be of actual mixtures. Then, the united strength, and not the isolated strength of a single ingredient, would be evaluated. It is not now the custom to make such determinations,—at least prior to the selection and use of materials, but it would be well so to do. This is a growing practice during the course of the work, as has been mentioned before, and it is a very commendable one. Its value is strikingly indicated by the results obtained by the Philadelphia Testing Laboratories, reported by W. Purvis Taylor and quoted in the discussion on tensile strength. It is there seen that a cement which on the pre-tensile determination appeared below standard gave actually the best results. With this fact before us, even though it be an isolated instance, it would seem advisable to include a pre-compressive determination on all the materials to be used and in the actual combination. The compressive strength of the *cement*, be it measured directly or indirectly by

---

*Chapter VI.

means of the tensile determination, is after all a relative, not an absolute measure of the compressive strength of the concrete.

Method of Molding Test Specimen:—In making test specimens for ascertaining directly the compressive strength, it is very important that the surface of the mold be suitable. A rough, dry, porous wooden mold is extremely undesirable. If wood is used it should be planed and the pores well sealed with oil, and at least 1¼ in. stuff, as lighter material warps too easily. Metal, or metal lined molds, are much better. The specimen should be removed as soon as possible after solidifying, generally in 24 hours, and either placed under water or covered with damp cloths, being kept in a moist condition until time for testing. It·is also necessary to cover the mold during the first 24 hours with wet cloths, being careful to see that they do not touch the soft material. In removing from the mold, great care needs be exercised not to mar the specimen; for it is still quite soft and punky and will be sure to suffer from rough treatment. This fact should be borne in mind in making the mold: It should be designed so as to be readily withdrawn from the concrete and not the concrete withdrawn from them; a type with collapsible sides is most suitable. Another very excellent mold is the sand mold, which is similar to the kind used in foundries. The sand acts to absorb the surplus moisture and at the same time to keep the specimen in a constant condition of moistness. Test pieces so cast are said to show the highest results.

Tests on Actual Mixture:—The most valuable indications as to compressive strength of actual mixtures are obtained by securing the material for the molds from the concrete as it is placed. Samples may be collected in a pail or other vessel from several parts of a day's installation after the concrete has been placed and tamped, taking a little here and a little there. At least three cubes should be secured for each days' work of each kind of concrete or mortar used; then tests can be made at 7 days, 30 days, and 60 days respectively. An endeavor should be made to surround these cubes while hardening with the same conditions of temperature and moistness as the work itself. Then, if the molding be carefully accomplished, exceedingly valuable indications can be secured as to the condition of the concrete in the work. Aside from the satisfaction of knowing at all times the strength of the concrete actually going into the work, the indications have an important practical value in that they furnish a fair guide as to the proper time to remove centers and put the structure into use. Both the cubes and the plans should be marked in accord, in order to definitely locate where the concrete was taken.*

SOUNDNESS OR CONSTANCY OF VOLUME:

Importance:—The *soundness* or *constancy-of-volume* is

---

*In this connection the methods used on the Harrisburg bridge, described in Concrete Engineering, September, 1909, are of interest.

the *most important*, as it is the *least determinate* of the properties of cement. It is most important for the reason that any unsoundness in a cement may be distinctly dangerous to the very integrity, the strength and durability, of the structure into which it enters, whereas in other respects inferiority may only make for impairment, but not virtual ruin. It is least determinate because up till the present time no satisfactory method has been evolved for making an absolute or nearly absolute determination, results frequently going by opposition rather than direction and cements evincing no signs of unsoundness under the ordinary tests, may yet contain the germs of ultimate dissolution.

Definition:—*Unsoundness* may best be described as the possession inherent in the cement—due for the most part to errors in manufacture—of unstable elements which in the work tend to swell and blow and thus to induce cracking and disintegration. Causes commonly assigned are:

(1) Improper chemical composition of raw materials.

(2) Insufficient reduction and amalgamation of raw materials during the operation of preliminary grinding.

(3) Incomplete calcination in the kiln.

(4) Predominance in finished product of small, unground particles of clinker.

(5) Adulteration with natural cement.

Perhaps the most specific source of unsoundness results from the presence of free or loosely combined lime. Ordinary free lime will hydrate very quickly upon the addition of water, but in the condition that it occurs, or may occur, in cements it hydrates much more slowly, probably because in a different chemical state and perhaps loosely combined with other elements. Hence, unless present in considerable quantity, there will be no violent manifestation of hydration, as in the ordinary case of slaking lime, but it may be so slow as at first to be imperceptible. This is the danger, for it will continue to hydrate within the hardened concrete, and in so doing exert an expansive effect upon the mass that may entirely destroy it. There is always liable to be more or less of this agency in freshly ground cements, in spite of the utmost perfection of the process of manufacture, as was remarked in the section on Manufacture, and manufacturers seek to reduce it by one or more of the following treatments:

(1) Steaming the clinker during the operation of fine grinding.

(2) Watering the clinker as it emerges from the kiln.

(3) Aerating the finished product before packing for shipment.

By combination of these, if not by any one alone, most of the objectionable element—unless due to radical error in the process of manufacture—may be rendered inoffensive. These same devices were mentioned under *"set"*, and it is probably for the same reason that they are effective for other purposes.

Effect of Time:—If the amount of free lime is slight—incidental rather than consequential—it may not materially impair the value of the product, although it may cause the cement, when freshly ground, to fail of passing some of the prescribed tests for soundness, particularly the "accelerated tests" (to be described). If other indications are favorable such failure should not necessarily occasion the rejection of the cement, for usually after a few weeks' aging all suggestion of unsoundness will have lapsed. The change that takes place is simple: it is the combination of the free lime with the moisture of the atmosphere to form slaked or hydrated lime, which in turn finally combines with the carbonic acid gas in the atmosphere to form lime carbonate (same as limestone). Lime hydrated is rather a benefit than otherwise. In fact it is often added to concrete to increase water-tightness, and in percentages up to 10 per cent seems to have no deteriorating effect upon the strength. It is also added to cement mortars, in amounts up to 25 per cent, with benefit, increasing the ease of spreading very materially, cheapening and not reducing the strength. *But in the free condition, as quicklime, it is a positive detriment.*

Relative Value of Indication:—If *unsoundness* is indicated on testing while all other determinations are satisfactory it is usually safe to conclude that the difficulty is not serious and will be remedied with age. If, however, it occurs in combination with other evidences of weakness or inferiority, it will constitute offhand sufficient warrant for rejection, indicating usually serious errors in process of manufacture.

Soundness Indication Alone Insufficient:—The simple determinations to detect unsoundness are seldom alone sufficient, inasmuch as they are neither absolute at the best nor quantitative ever,—that is, the extent and amount of the deleterious elements are not hereby revealed,—hence they are only *positive* when *rightly interpreted* in conjunction with the results of other determinations of quality.

Presence of Unground Clinker:—*Unsoundness*, due to the predominance in the finished cement of unground particles of clinker, is not so serious as unsoundness due to incomplete or incorrect composition of the clinker, indicating, as the latter does, fundamental errors. Such condition may not materially affect the integrity of the structure. The presence of such a cause would be disclosed positively by the determination of fineness and tensile strength.

Tests for Soundness:—The principal methods for determining soundness, in vogue at the present time, are: (1) The cold water test; (2) the hot water test; (3) the boiling water test; and (4) the steam test. Of these the first is the most commonly accepted standard; the remaining are in reality developments or modifications of it, devised primarily for hastening the determination, hence are characterized as "accelerated tests." The requirements for first-quality cements are that pats of it neat, hardened first in moist air for 24 hours, shall under water at the varying degrees of temperature represented by the several tests exhibit no signs of swelling, checking, cracking, distortion or disintegration, remaining throughout firm, hard and true. The cement is then said to be *sound*, or have a *constant-volume*, and is regarded as acceptable for usage. Failure to pass the accelerated tests is not generally accepted as sufficient warrant for condemnation, on account of their relative severity over practical working conditions, but failures of the test-pats to remain perfect under water at 70 degrees F., in from seven to 14 days, is held to strongly indicate the unfitness of the material for use.

The Normal Test:—The cold-water test is usually characterized as the "Normal Test." The method is simply to immerse a 24-hour old pat of neat cement, on a glass plate, in water at 70 degrees F., and to observe it throughout a period of 28 days. A similar pat is for comparison observed simultaneously in moist air. If both remain firm and hard and true the cement is pronounced *sound*. But signs of cracking, distortion and disintegration are indications of *unsoundness*, and held as warrant for the rejection.

The Accelerated Tests:—The accelerated tests were devised primarily to hasten the determination, based upon the fact that the higher the temperature the more rapid the chemical action which is involved in the hardening of cement. Thus, in a few hours it is possible to secure the same effect as in 28 days normally. The action is, however, relatively more severe, especially the boiling water and steam tests, so severe in fact that they are not yet accepted as criteria for the rejection of a cement, although most standard brands will pass them safely. Their greatest value, perhaps, has been to raise the standard of manufacture, to make manufacturers more particular as to the quality of their product.

The steam test, due to M. Le Chatelier, is the most complete test, as only virtually a perfectly combined cement will pass it. Insufficient reduction of raw materials and amalgamation of same, and incomplete or improper calcination, even in small degree, are hereby revealed with certainty. The ordeal is, however, too severe as a *practical* criterion, for it will develop *unsoundness* in cements that, under working conditions, will give perfect satisfaction. There is one situation where it is of practical worth, namely, for cement intended for extremely hot and humid usage, as e. g., tanks in which hot solutions may be contained.

An Expert's Opinion:—Mr. W. Purvis Taylor, in charge of the testing laboratories of the city of Philadelphia, in a paper read before the Association of American Portland Cement Manufacturers, had this to say about the question of soundness (December, 1904):

"The soundness tests are undoubtedly the most interesting of the physical tests of cement, not only from the importance of the subject, but also from the many contradictions that develop in their study. Almost every day we seem to have a case that upsets all the opinions one has previously formed. We cannot but admit that at present we *do not know how to test cement for soundness.* The microscope may ultimately solve some of these questions for us, but at present that branch of study is too embryonic to be of much practical value. *The only infallible test of soundness is that of the work itself, but as a test this is rather unsatisfactory.* The nearest approach to this is, of course, the normal pat tests, but their indications require such a long time as to make them almost valueless as acceptance tests. We have on record cases where soundness developed in normal pats only after five years, and it would scarcely be convenient to hold cement under test that long. * * * *

"* * * * Regarding the comparison of soundness tests with other physical tests and with actual construction we still have the same hopeless maze of contradictions and irregularities with which we are all familiar."

Commenting upon this extremely valuable opinion, coming as it does from one so long experienced in the art of cement testing, while in no way detracting from its worth, it should be remembered that it comes also from one who is active in the theoretical side of the question, hence inclined to be very critical, and that on a practical plane things are by no means so indefinite.

Value of Test Data:—It is evident, however, that the properties of cement, although very important to know, are not at the present time subject to precise determination, placed on the results of test data. "The proof of the pudding is in the eating," and the comparatively insignificant number of failures of concrete work assignable to inferior cement in comparison to the magnificent total of examplar structures are everlasting monuments of the efficiency of Portland cement. This much may be said of the determinations: they are on the average sufficiently accurate to safeguard against radically bad product and are hence by no means to be lightly regarded.

PURITY OF THE CEMENT:

Impurities in Cement Are of Two Kinds:—(1) Those incidental to the process of manufacture and for the most part unavoidable, and (2) intentional adulteration. The proclivity of the human to adulterate and substitute in an endeavor to cheapen the cost of production at the expense of the unwary consumer, is occasionally met with in connection with the cement industry, as it is in every walk of life. The two commonest methods of adulteration are:

(1) The blend with Natural cement.

(2) To intermingle finely-ground inerts, such as slag, limestone, siliceous sand, dust, etc.

Adulteration With Natural Cement:—This is liable to be expected only of mills that manufacture both kinds of cement—Natural and Portland. Of such there are not, fortunately, at the present time, many examples, as it hardly pays a mill to bother with the two processes. This adulteration may or may not be injurious, depending on the quality of the interblended Natural cement. It will usually show in a lighter specific gravity and a lower rank in fineness in any case, and if an inferior Natural, in both a reduced tensile strength and unsoundness. The effect of blending an unsound Natural with a sound Portland is well illustrated in the following table (credit given Mr. W. Purvis Taylor):

Per cent Portland Cement.	Per cent Natural Cement.	Tests for Soundness.			
		Boiling.	Steam.	Water 185° F.	Water 140° F.
100	0	O. K.	O. K.	O. K.	O. K.
90	10	O. K.	O. K.	O. K.	O. K.
80	20	O. K.	O. K.	O. K.	O. K.
70	30	O. K.	O. K.	O. K.	O. K.
60	40	Sl. Ck.	O. K.	O. K.	O. K.
50	50	Ck. & Crack.	O. K.	O. K.	O. K.
40	60	Bad Ck.	Sl. Ck.	Sl. Ck.	O. K.
30	70	Disint.	Ck.	Sl. Ck.	O. K.
20	80	Disint.	Ck. & Crack.	Ck.	Sl. Ck.
10	90	Disint.	Part disint.	Part disint.	Sl. Ck.
0	100	Disint.	Disint.	Disint.	Ck. & Crack.

The Acid Test:—The detection of adulteration may be very simply and conveniently determined by treating a small sample of the cement, about as much as can be piled on a 5-cent piece, with muriatic acid. A test-tube should be used for this operation, and the action facilitated by stirring with a glass rod. Excessive lime will be indicated by marked effervescence. Adulteration with Natural cement by the throwing down of a bright yellow jelly; the presence of underburnt particles by the same sign; addition of foreign matter, like ground slag and silica, subsequent to calcination, by separation out as a fine sediment in the tube. The pure cement will also be converted into a jelly, but of orange hue. If not all the silica is combined it will also be separated out by the acid and show as so much insoluble fine grit—harmless to the cement except for that it reduces the sand-carrying capacity slightly, acting as so much inert matter.

Adulteration With Pulverized Grits:—Harmless so far as the quality of soundness goes, but detrimental to the strength, acting as so much fine sand. Usually added to the clinker and so ground right into the product, which makes it difficult to discern. It is a convenient method of adulteration for the sake of cheapening, hence liable to be common. It may be detected by the specific gravity determination, making for lighter weight, the tensile strength, which it lowers, and the acid test described in (b). It may or may not be disclosed in the sifting for fineness, the grains of grit being usually as fine as the grains of cement. It may also be determined, but not infallibly, depending on the material used as an adulterant, by chemical analysis. There are cases where this adulteration is purposed, and the product is honestly marketed as a "siliceous-cement," or some such, and for less important usages it has a legitimate application; but never for *structural-concrete*. The limit of material allowed to be added after the burning is fixed by Standard Specifications at 3 per cent, hence if 2 per cent of this be taken up by set-restrainer, there is no margin for other adulteration in a bona fide Portland cement.

Lime:—Free lime is usually present, not by intent, for its deleterious effect is too well recognized and every manufacturer tries to get rid of it, but by accident or fault in the composition of materials or process of manufacture. If lime is added by intent it is always in the hydrated form, in which it is rather beneficial than harmful, improving the setting qualities. Its percentage should, however, be limited strictly. It is not a recognized constituent of normal Portland cements, and if added at all should be added by the consumer as he uses it and for a particular purpose. It will be revealed, no matter what form, free or hydrated, by the acid test by effervescence, and in the accelerated tests for soundness, in the free form only, by swelling and cracking of the test-specimen.

The ordinary quicklime, which is burned at a low red heat, finally ground and mixed with water, will slake very quickly, hence would be hardly likely to injure concrete, being hydrated during the preliminary stages. Lime which has been burned at the white heat of a cement kiln seems to act contrarywise, slaking extremely slow, probably because in a loosely combined state, hence its affinity with water may continue after the concrete is in place and hardened. The result is swelling and possible disintegration of the mass, and certainly its unsightly discoloration.

The action of lime in this form, in case the concrete is exposed to the air, may be so delayed as not to manifest expansion for several months or longer, but if under water or under moist conditions continually the action will be much more prompt, so that if there is going to be any injurious effect it will manifest itself within a few weeks.

The vital importance of having a cement with little or no incompletely combined lime is obvious.

Magnesia:—*Free magnesia* is present unavoidably, and in small percentages is not considered injurious. The amount is limited by *Standard Specifications* to 4 per cent. There is no good way of detecting its presence, except by chemical analysis. Its effect on the cement is similar to that of lime, only more insidious, as its hydration and hence its expansive tendency is more gradual. The injurious effects of lime will be manifested within a few weeks after using, but the ultimate effect of magnesia may not be felt for several years and be so slow and gradual as to be irresistible in its force.

Lime is usually revealed in the accelerated tests for soundness, but magnesia, except in exceptionally excessive amounts, will not be. Insomuch as its hydration is extremely slow, as compared to lime, there is little danger to be feared except where the concrete is exposed more or less of the time to water, and if the percentage is no greater than the limit specified there is little likelihood of danger anyhow.

Sulphur:—*Sulphur*, like magnesia, is one of the tabooed elements in cement, the *Standard Specifications* limiting it to 1.75 per cent (in the acid-radical form, SO_3). The danger is this—that, being in the acid-radical form it is prime to combine with water to form sulphuric acid, H_2SO_4, which will attack cement. Thus, in any considerable amounts in the body of a mass of cement, or concrete, it would result in the formation of acid, hence "rotten at the core." In the reaction sulphuretted-hydrogen, H_2S, would be liberated, which, seeking an outlet, would exert an expansive force, disintegrating the concrete. This is the principal objection to the use of slag cement, and to aggregates like slag and cinders that contain, or are likely to, quantities of sulphur. The author knows of a case where a mass of concrete made with slag-aggregates, evincing signs of unsoundness, was split open and found to be very nearly disintegrated. The cause was quite apparent by the strong odor of "decayed eggs" that was therewith released, which is the well-known odor of H_2S.

Fortunately, in Portland cement, there is little danger on this score, since virtually all sulphides in the raw materials are ignited and driven off as fumes up the flue.

Gypsum and Calcium Chloride are other adulterants liable to be present, being usually added by design for the purpose of regulating the set. They are—in small per cents—comparatively harmless and need not be tested for.

Substitution:—Because a cement comes in a standard-brand sack is no proof positive that it is the brand indicated. Dealers have been known to repack sacks of a high-class brand with inferior material and slip it in on the unsuspecting purchaser. When the product is bought of the mill direct—as is the case in large lots—there is little likelihood of meeting with this fraud. It is the small consumer who is forced to purchase from the middleman who needs be on the alert.

Chemical Composition:—The chemical composition is important, but concerns the user less than the manufacturer; it is seldom reported unless the results of the other determinations seem to require it. It is used principally for detecting the proportions, etc., in ingredients believed in excess of certain percentages to be harmful: that is, SO_3 and MgO. It is also valuable in detecting the adulteration of the cement with considerable amounts of inert material, such as slag, ground limestone or siliceous sand. The determination of the principal constituents of cement is interesting, but not conclusive, as an indication of the quality of the finished product, since errors for the most part occur, not in the composition of the raw materials, but in their mixing and calcination. The indication of the amount of lime affords little clue also, inasmuch as a high-limed cement, by reason of finer grinding and harder burning, may be superior to a low-limed cement poorly ground and carelessly burned; the latter and not the former may then be the dangerous cement. So, too, the ash of the fuel used in burning may so alter the composition of the finished product as to vitiate the value of the chemical analysis. Taken in conjunction with the other determinations, it may explain things not otherwise clear. It is a determination for the analytical chemist, and for sample analyses and more specific explanation reference is made to the chapter on Chemical Composition.

There is one instance where the test of chemical composition is of essential practical value,—namely, in case the cement is intended for sea-water use, where it has been shown that cements high in alumina are unadapted. When, therefore, the cement is intended for such purposes the chemical analysis should always be made. For safe results the cement should be as near the ideal composition as possible, any especial variance in composition or manufacture, aside from abnormal alumina content, which might not debar the cement from ordinary work, being here cause for rejection.

STANDARD SPECIFICATIONS FOR PORTLAND CEMENT.*

Definition:—This term is applied to the finely pulverized product resulting from the calcination to incipient fusion of an intimate mixture of properly proportioned argillaceous and calcareous materials, and to which no addition greater than 3 per cent has been made subsequent to calcination.

* From "Standard Methods of Testing and Specifications for Cement," edited under the direction of Committee C of the American Society for Testing Materials. Published by the Committee, 1909.

Specific Gravity:—The specific gravity of the cement ignited at a low red heat shall not be less than 3.10; and the cement shall not show a loss on ignition of more than 4 per cent.

Fineness:—It shall leave by weight a residue of not more than 8 per cent on the No. 100, and not more than 25 per cent on the No. 200 sieve.

Time of Setting:—It shall not develop initial set in less than 30 minutes; and must develop hard set in not less than one hour, nor more than ten hours.

Tensile Strength:—The minimum requirements for tensile strength for briquettes 1 in. sq. in section shall be within the following limits, and shall show no retrogression in strength within the periods specified:*

Age	Neat Cement.	Strength.
24 hours in moist air		150-200 lb.
7 days (1 day in moist air, 6 days in water)		450-550 lb.
28 days (1 day in moist air, 27 days in water)		550-650 lb.

One Part Cement, Three Parts Sand.

7 days (1 day in moist air, 6 days in water)		150-200 lb.
28 days (1 day in moist air, 27 days in water)		†200-300 lb.

Constancy of Volume:—Pats of neat cement about 3 in. in diameter, ½-in. thick at the center, and tapering to a thin edge, shall be kept in moist air for a period of 24 hours.

(a) A pat is then kept in air at normal temperature and observed at intervals for at least 28 days.

(b) Another pat is kept in water maintained as near 70 degrees F. as practicable, and observed at intervals for at least 28 days.

(c) A third pat is exposed in any convenient way in an atmosphere of steam, above boiling water, in a loosely closed vessel for five hours.

These pats, to satisfactorily pass the requirements, shall remain firm and hard and show no signs of distortion, checking, cracking or disintegrating.

Sulphuric Acid and Magnesia:—The cement shall not contain more than 1.75 per cent of anhydrous sulphuric acid (SO₄), nor more than 4 per cent of magnesia (MgO).

General Observations:—These remarks have been prepared with a view of pointing out the pertinent features of the various requirements and the precautions to be observed in the interpretation of the results of the tests.

The committee would suggest that the acceptance or rejection under these specifications be based on tests made by an experienced person having the proper means for making the tests.

*For example, the minimum requirement for the 24 hour heat cement test should be some specified value within the limits of 150 and 200 lbs., and so on for each period stated.

†If the minimum strength is not specified the mean of the values shall be taken as the minimum strength required.

Specific Gravity:

Specific gravity is useful in detecting adulteration. The results of tests of specific gravity are not necessarily conclusive as an indication of the quality of a cement, but when in combination with the results of other tests may afford valuable indications.

Fineness:

The sieves should be kept thoroughly dry.

Time of Setting:

Great care should be exercised to maintain the test pieces under as uniform conditions as possible. A sudden change or wide range of temperature in the room in which the tests are made, a very dry or humid atmosphere, and other irregularities vitally affect the rate of setting.

Tensile Strength:

Each consumer must fix the minimum requirements for tensile strength to suit his own conditions. They shall, however, be within the limits stated.

Constancy of Volume:

The tests for constancy of volume are divided into two classes, the first normal, the second accelerated. The latter should be regarded as a precautionary test only, and not infallible. So many conditions enter into the making and interpreting of it that it should be used with extreme care.

In making the pats the greatest care should be exercised to avoid initial strains due to molding or to too rapid drying-out during the first 24 hours. The pats should be preserved under the most uniform conditions possible, and rapid changes of temperature should be avoided.

The failure to meet the requirements of the accelerated tests need not be sufficient cause for rejection. The cement may, however, be held for 28 days, and a retest made at the end of that period, using a new sample. Failure to meet the requirements at this time should be considered sufficient cause for rejection, although in the present state of our knowledge it cannot be said that such failure necessarily indicates unsoundness, nor can the cement be considered entirely satisfactory simply because it passes the tests.

General conditions:—

1. All cement shall be inspected.

2. Cement may be inspected either at the place of manufacture or on the work.

3. In order to allow ample time for inspecting and testing, the cement should be stored in a suitable weather-tight building having the floor properly blocked or raised from the ground.

4. The cement should be stored in such a manner as to permit easy access for proper inspection and identification of each shipment.

5. Every facility shall be provided by the contractor and a period of at least 12 days allowed for the inspection and necessary tests.

6. Cement shall be delivered in suitable packages with the brand and name of manufacturer plainly marked thereon.

7. A bag of cement shall contain 94 pounds of cement net. Each barrel of Portland cement shall contain four bags, and each barrel of Natural cement shall contain three bags of the above net weight.

8. Cement failing to meet the seven-day requirements may be held awaiting the results of the 28-day tests before rejection.

9. All tests shall be made in accordance with the methods proposed by the Committee on Uniform Tests of Cement of the American Society of Civil Engineers, presented to the Society Jan. 21, 1903, and amended Jan. 20, 1904, and Jan. 14, 1908, with all subsequent amendments thereto.

Standard Methods of Testing

Sampling:—

1. Selection of Sample.—The selection of the sample for testing is a detail that must be left to the discretion of the engineer, the number and the quantity to be taken from each package will depend largely on the importance of the work, the number of tests to be made and the facilities for making them.

2. The sample shall be a fair average of the contents of the package; it is recommended that, where conditions permit, one barrel in every ten be sampled.

3. Samples should be passed through a sieve having 20 meshes per linear inch, in order to break up lumps and remove foreign material; this is also a very effective method for mixing them together in order to obtain an average. For determining the characteristics of a shipment of cement, the individual samples may be mixed and the average tested; where time will permit, however, it is recommended that they be tested separately.

4. Method of Sampling.—Cement in barrels should be sampled through a hole made in the center of one of the staves, midway between the heads, or in the head, by means of an auger or a sampling iron similar to that used by sugar inspectors. If in bags, it should be taken from surface to center.

Chemical Analysis:—

5. Significance.—Chemical analysis may render valuable service in the detection of adulteration of cement with considerable amounts of inert material, such as slag or ground limestone. It is of use, also, in determining whether certain constituents, believed to be harmful when in excess of a certain per-

centage, as magnesia and sulphuric anhydride, are present in inadmissible proportions.

6. The determination of the principal constituents of cement—silica, alumina, iron oxide and lime—is not conclusive as an indication of quality. Faulty character of cement results more frequently from imperfect preparation of the raw material or defective burning than from incorrect proportions of the constituents. Cement made from very finely-ground material, and thoroughly burned, may contain much more lime than the amount usually present, and still be perfectly sound. On the other hand, cements low in lime may, on account of careless preparation of the raw material, be of dangerous character. Further, the ash of the fuel used in burning may so greatly modify the composition of the product as largely to destroy the significance of the results of analysis.

7. Method.—As a method to be followed for the analysis of cement, that proposed by the Committee on Uniformity in the Analysis of Materials for the Portland Cement Industry, of the New York Section of the Society for Chemical Industry, and published in *Engineering News*, Vol. 50, p. 60, 1903; and in *The Engineering Record*, Vol. 48, p. 49, 1903, is recommended.

Specific Gravity:—

8. Significance.—The specific gravity of cement is lowered by adulteration and hydration, but the adulteration must be in considerable quantity to affect the results appreciably.

9. Inasmuch as the differences in specific gravity are usually very small, great care must be exercised in making the determination.

10. Apparatus and Method.—The determination of specific gravity is most conveniently made with Le Chatelier's apparatus. This consists of a flask (D), Fig. 45, of 120 cu. cm. (7.32 cu. in.) capacity, the neck of which is about 20 cm. (7.87 in.) long; in the middle of this neck is a bulb (C), above and below which are two marks (F) and (E); the volume between these marks is 20 cu. cm. (1.22 cu. in.) The neck has a diameter of about 9 mm. (0.35 in.), and is graduated into tenths of cubic centimeters above the mark (F).

11. Benzine (62 deg. Baume naphtha), or kerosene free from water, should be used in making the determination.

12. The specific gravity can be determined in two ways:

(1) The flask is filled with either of these liquids to the lower mark (E), and 64 g. (2.25 oz.) of powder, cooled to the temperature of the liquid, is gradually introduced through the funnel (B) [the stem of which extends into the flask to the top of the bulb (C)], until the upper mark (F) is reached. The difference in weight between the cement remaining and the original quantity (64 g.) is the weight which has displaced 20 cu. cm.

13. (2) The whole quantity of the powder is introduced, and

181

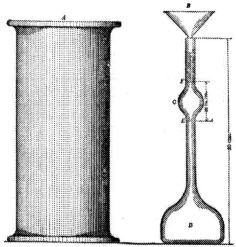

FIG. 45—LE CHATELIER'S SPECIFIC GRAVITY APPARATUS.

the level of the liquid rises to some division of the graduated neck.
This reading plus 20 cu. cm. is the volume displaced by 64 g. of
the powder.

14. The specific gravity is then obtained from the formula:

$$\text{Specific Gravity} = \frac{\text{Weight of Cement, in grammes,}}{\text{Displaced Volume, in cubic centimeters.}}$$

15. The flask, during the operation, is kept immersed in
water in a jar (A), in order to avoid variations in the temperature of the liquid. The results should agree within 0.01. The
determination of specific gravity should be made on the cement
as received; and, should it fall below 3.10, a second determination
should be made on the sample ignited at a low red heat.

16. A convenient method for cleaning the apparatus is as
follows: The flask is inverted over a large vessel, preferably a
glass jar, and shaken vertically until the liquid starts to flow
freely; it is then held still in a vertical position until empty;
the remaining traces of cement can be removed in a similar manner by pouring into the flask a small quantity of clean liquid and
repeating the operation.

17. More accurate determinations may be made with the
picnometer.

Fineness:—

18. Significance.—It is generally accepted that the coarser
particles in cement are practically inert, and it is only the ex-

tremely fine powder that possesses adhesive or cementing qualities. The more finely cement is pulverized, all other conditions being the same, the more sand it will carry and produce a mortar of a given strength.

19. The degree of final pulverization which the cement receives at the place of manufacture is ascertained by measuring the residue retained on certain sieves. Those known as the No. 100 and No. 200 sieves are recommended for this purpose.

20. Apparatus.—The sieves should be circular, about 20 cm. (7.87 in.) in diameter, 6 cm. (2.36 in.) high, and provided with a pan 5 cm. (1.97 in.) deep, and a cover.

21. The wire cloth should be of brass wire having the following diameters:

No. 100. 0.0045 in.; No. 200, 0.0024 in.

22. This cloth should be mounted on the frames without distortion; the mesh should be regular in spacing and be within the following limits:

No. 100, 96 to 100 meshes to the linear inch.
No. 200, 188 to 200 meshes to the linear inch.

23. Fifty grammes (1.76 oz.) or 100 g. (3.52 oz.) should be used for this test, and dried at a temperature of 100 degrees C. (212 degrees F.) prior to sieving.

24. Method.—The Committee, after careful investigation, has reached the conclusion that mechanical sieving is not as practicable or efficient as hand work, and therefore recommends the following method:

25. The thoroughly dried and coarsely screened sample is weighed and placed on the No. 200 sieve, which, with pan and cover attached, is held in one hand in a slightly inclined position, and moved forward and backward, at the same time striking the side gently with the palm of the other hand, at the rate of about 200 strokes per minute. The operation is continued until not more than one-tenth of 1 per cent passes through after one minute of continuous sieving. The residue is weighed, then placed on the No. 100 sieve and the operating repeated. The work may be expedited by placing in the sieve a small quantity of large steel shot. The results should be reported to the nearest tenth of 1 per cent.

Normal Consistency:—

26. Significance.—The use of a proper percentage of water in making the pastes* from which pats, tests of setting, and briquettes are made, is exceedingly important, and affects vitally the results obtained.

27. The determination consists in measuring the amount of water required to reduce the cement to a given state of plasticity, or to what is usually designated the normal consistency.

*The term "paste" is used in this report to designate a mixture of cement and water, and the word "mortar" a mixture of cement, sand and water.

28. Various methods have been proposed for making this determination, none of which has been found entirely satisfactory. The Committee recommends the following:

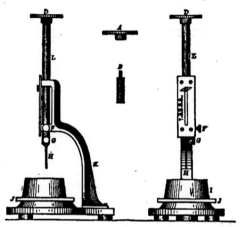

FIG. 46—VICAT NEEDLE.

29. Method. Vicat Needle Apparatus.—This consists of a frame (*K*), Fig. 46, bearing a movable rod (*L*), with the cap (*A*) at one end, and at the other the cylinder (*B*), 1 cm. (0.39 in.) in diameter, the cap, rod, and cylinder weighing 300 g. (10.58 oz.). The rod, which can be held in any desired position by a screw (*F*), carries an indicator, which moves over a scale (graduated to centimeters) attached to the frame (*K*). The paste is held by a conical, hard-rubber ring (*I*), 7 cm. (2.76 in.) in diameter at the base, 4 cm. (1.57 in.) high, resting on a glass plate (*J*), about 10 cm. (3.94 in.) square.

30. In making the determination, the same quantity of cement as will be subsequently used for each batch in making the briquettes, but not less than 500 g., is kneaded into a paste, as described in Paragraph 56, and quickly formed into a ball with the hands, completing the operation by tossing it six times from one hand to the other, maintained 6 in. apart; the ball is then pressed into the rubber ring, through the larger opening, smoothed off, and placed (on its large end) on a glass plate and the smaller end smoothed off with a trowel; the paste, confined in the ring, resting on the plate, is placed under the rod bearing the cylinder which is brought in contact with the surface and quickly released.

31. The paste is of normal consistency when the cylinder

penetrates to a point in the mass 10 mm. (0.39 in.) below the top of the ring. Great care must be taken to fill the ring exactly to the top.

32. The trial pastes are made with varying percentages of water until the correct consistency is obtained.

• 33. The Committee has recommended, as normal, a paste, the consistency of which is rather wet, because it believes that variations in the amount of compression to which the briquette is subjected in moulding are likely to be less with such a paste.

34. Having determined in this manner the proper percentage of water required to produce a paste of normal consistency, the proper percentage required for the mortars is obtained from an empirical formula.

35. The Committee hopes to devise such a formula. The subject proves to be a very difficult one, and although the Committee has given it much study, it is not yet prepared to make a definite recommendation.

NOTE. The Committee on Standard Specifications for Cement inserts the following table for temporary use to be replaced by one to be devised by the Committee of the American Society of Civil Engineers.

Neat.	One Cement Three Standard Ottawa Sand.	Neat.	One Cement Three Standard Ottawa Sand.	Neat.	One Cement Three Standard Ottawa Sand.
15	8.0	23	9.3	31	10.7
16	8.2	24	9.5	32	10.8
17	8.3	25	9.7	33	11.0
18	8.5	26	9.8	34	11.2
19	8.7	27	10.0	35	11.5
20	8.8	28	10.2	36	11.5
21	9.0	29	10.3	37	11.7
22	9.2	30	10.5	38	11.8

	1 to 1	1 to 2	1 to 3	1 to 4	1 to 5
Cement	500	333	250	200	167
Sand	500	666	750	800	833

Time of Setting:—

36. Significance.—The object of this test is to determine the time which elapses from the moment water is added until the paste ceases to be fluid and plastic (called the "initial set"), and also the time required for it to acquire a certain degree of hardness (called the "final" or "hard set"). The former of these is the more important, since, with the commencement of setting, the process of crystallization or hardening is said to begin. As a disturbance of this process may produce a loss of strength, it is desirable to complete the operation of mixing and moulding

or incorporating the mortar into the work before the cement begins to set.

37. It is usual to measure arbitrarily the beginning and end of the setting by the penetration of weighted wires of given diameters.

38. Method.—For this purpose the Vicat Needle, which has already been described in Paragraph 29, should be used.

39. In making the test, a paste of normal consistency is moulded and placed under the (L), Fig. 46, as described in Paragraph 30; this rod, bearing the cap (D) at one end and the needle (H), 1 mm. (0.039 in.) in diameter, at the other, weighing 300 g. (10.58 oz.). The needle is then carefully brought in contact with the surface of the paste and quickly released.

40. The setting is said to have commenced when the needle ceases to pass a point 5 mm. (0.20 in.) above the upper surface of the glass plate, and is said to have terminated the moment the needle does not sink visibly into the mass.

41. The test pieces should be stored in moist air during the test; this is accomplished by placing them on a rack over water contained in a pan and covered with a damp cloth, the cloth to be kept away from them by means of a wire screen; or they may be stored in a moist box or closet.

42. Care should be taken to keep the needle clean, as the collection of cement on the sides of the needle retards the penetration, while cement on the point reduces the area and tends to increase the penetration.

43. The determination of the time of setting is only approximate, being materially affected by the temperature of the mixing water, the temperature and humidity of the air during the test, the percentage of water used, and the amount of moulding the paste receives.

Standard Sand.

44. The Committee recognizes the grave objections to the standard quartz now generally used, especially on account of its high percentage of voids, the difficulty of compacting in the moulds, and its lack of uniformity; it has spent much time in investigating the various natural sands which appeared to be available and suitable for use.

45. For the present, the Committee recommends the natural sand from Ottawa, Ill., screened to pass a sieve having 20 meshes per linear inch and retained on a sieve having 30 meshes per linear inch; the wires to have diameters of 0.0165 and 0.0112 in., respectively, i. e., half the width of the opening in each case. Sand having passed the No. 20 sieve shall be considered standard when not more than 1 per cent passes a No. 30 sieve after one minute's continuous sifting of a 500 g. sample.*

*The Sandusky Portland Cement Company of Sandusky, O., has agreed to undertake the preparation of this sand, and to furnish it at a price only sufficient to cover the actual cost of preparation.

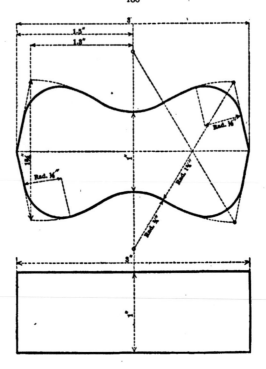

FIG. 47—DETAIL OF STANDARD BRIQUETTE.

Form of Briquette:—

46. While the form of the briquette recommended by a former Committee of the Society is not wholly satisfactory, this Committee is not prepared to suggest any change, other than rounding off the corners by curves of ½-in. radius, Fig. 47.

Moulds:—

47. The moulds should be made of brass, bronze, or some equally non-corrodible material, having sufficient metal in the sides to prevent spreading during moulding.

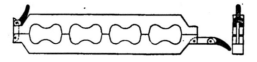

FIG. 48—DETAILS FOR GANG MOLD.

48. Gang moulds, which permit moulding a number of briquettes at one time, are preferred. by many to single moulds; since the greater quantity of mortar that can be mixed tends to produce greater uniformity in the results. The type shown in Fig. 48 is recommended.

49. The moulds should be wiped with an oily cloth before using.

Mixing:—

50. All proportions should be stated by weight; the quantity of water to be used should be stated as a percentage of the dry material.

51. The metric system is recommended because of the convenient relation of the gramme and the cubic centimeter.

52. The temperature of the room and. the mixing water should be as near 21 degrees C. (70 degrees F.) as it is practicable to maintain it.

53. The sand and cement should be thoroughly mixed dry. The mixing should be done on some non-absorbing surface, preferably plate glass. If the mixing must be done on an absorbing surface it should be thoroughly dampened prior to use.

54. The quantity of material to be mixed at one time depends on the number of test pieces to be made; about 1,000 g. (35.28 oz.) makes a convenient quantity to mix, especially by hand methods.

55. The Commitee, after investigation of the various mechanical mixing machines, has decided not to recommend any machine that has thus far been devised, for the following reasons:

(1) The tendency of most cement is to "ball up" in the machine, thereby preventing the working of it into a homogeneous paste; (2) there are no means of ascertaining when the mixing is complete without stopping the machine, and (3) the difficulty of keeping the machine clean.

56. Method.—The material is weighed and placed on the mixing table, and a crater formed in the center, into which the proper percentage of clean water is poured; the material on the outer edge is turned into the crater by the aid of a trowel. As soon as the water has been absorbed, which should not require more than one minute, the operation is completed by vigorously kneading with the hands for an additional one minute, the process being similar to that used in kneading dough. A sand-glass affords a convenient guide for the time of kneading. During the operation of mixing, the hands should be protected by gloves, preferably of rubber.

Moulding:—

57. Having worked the paste or mortar to the proper consistency, it is at once placed in the moulds by hand.

58. The Committee has been unable to secure satisfactory

results with the present moulding machines; the operation of machine moulding is very slow, and the present types permit of moulding but one briquette at a time, and are not practicable with the pastes or mortars herein recommended.

59. Method.—The moulds should be filled immediately after the mixing is completed, the material pressed in firmly with the fingers and smoothed off with a trowel without mechanical ramming; the material should be heaped up on the upper surface of the mould, and, in smoothing off, the trowel should be drawn over the mould in such a manner as to exert moderate pressure on the excess material. The mould should be turned over and the operation repeated.

60. A check upon the uniformity of the mixing and moulding is afforded by weighing the briquettes just prior to immersion, or upon removal from the moist closet. Briquettes which vary in weight more than 3 per cent from the average should not be tested.

Storage of the Test Pieces:—

61. During the first twenty-four hours after moulding, the test pieces should be kept in moist air to prevent them from drying out.

62. A moist closet or chamber is so easily devised that the use of the damp cloth should be abandoned if possible. Covering the test pieces with a damp cloth is objectionable, as commonly used, because the cloth may dry out unequally, and, in consequence, the test pieces are not all maintained under the same condition. Where a moist closet is not available, a cloth may be used and kept uniformly wet by immersing the ends in water. It should be kept from direct contact with the test pieces by means of a wire screen or some similar arrangement.

63. A moist closet consists of a soapstone or slate box, or a metal-lined wooden box—the lining being covered with felt and this felt kept wet. The bottom of the box is so constructed as to hold water, and the sides are provided with cleats for holding glass shelves on which to place the briquettes. Care should be taken to keep the air in the closet uniformly moist.

64. After twenty-four hours in moist air, the test pieces for longer periods of time should be immersed in water maintained as near 21 degrees C. (70 degrees F.) as practicable; they may be stored in tanks or pans, which should be of non-corrodible material.

Tensile Strength:—

65. The tests may be made on any standard machine. A solid metal clip, as shown in Fig. 50, is recommended. This clip is to be used without cushioning at the points of contact with the test specimen. The bearing at each point of contact should be

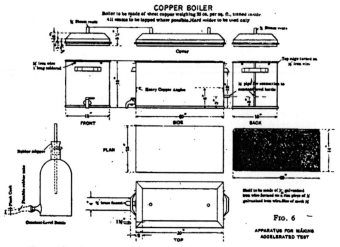

FIG. 49—APPARATUS FOR MAKING ACCELERATED TEST.

¼ in. wide, and the distance between the center of contact on the same clip should be 1¼ in.

66. Test pieces should be broken as soon as they are removed from the water. Care should be observed in centering the briquettes in the testing machine, as cross-strains, produced by improper centering, tend to lower the breaking strength. The load should not be applied too suddenly, as it may produce vibration, the shock from which often breaks the briquette before the ultimate strength is reached. Care must be taken that the clips and the sides of the briquette be clean and free from grains of sand or dirt, which would prevent a good bearing. The load should be applied at the rate of 600 lb. per min. The average of the briquettes of each sample tested should be taken as the test, excluding any results which are manifestly faulty.

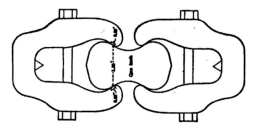

FIG. 50—DETAIL SHOWING FORM OF CLIP.

Constancy of Volume:—

67. Significance.—The object is to develop those qualities which tend to destory the strength and durability of a cement. As it is highly essential to determine such qualities at once, tests of this character are for the most part made in a very short time, and are known, therefore, as accelerated tests. Failure is revealed by cracking, checking, swelling, or disintegration, or all of these phenomena. A cement which remains perfectly sound is said to be of constant volume.

68. Methods.—Tests for constancy of volume are divided into two classes: (1) normal tests, or those made in either air or water maintained at about 21 degrees C. (70 degrees F.) and (2) accelerated tests, or those made in air, steam, or water at a temperature of 45 degrees C. (113 degrees F.) and upward. The test pieces should be allowed to remain 24 hours in moist air before immersion in water or steam, or preservation in air.

69. For these tests, pats, about 7½ cm. (2.95 in.) in diameter, 1¼ cm. (0.49 in.) thick at the center, and tapering to a thin edge, should be made, upon a clean glass plate [about 10 cm. (3.94 in.) square], from cement paste of normal consistency.

70. Normal Test.—A pat is immersed in water maintained as near 21 degrees C. (70 degrees F.) as possible for 28 days, and observed at intervals. A similar pat, after 24 hours in moist air, is maintained in air at ordinary temperature and observed at intervals.

71. Accelerated Test.—A pat is exposed in any convenient way in an atmosphere of steam, above boiling water, in a loosely closed vessel, for 5 hours. The apparatus recommended for making these determinations is shown in Fig. 6.

72. To pass these tests satisfactorily, the pats should remain firm and hard, and show no signs of cracking, distortion or disintegration.

73. Should the pat leave the plate, distortion may be detected best with a straight-edge applied to the surface which was in contact with the plate.

74. In the present state of our knowledge it cannot be said that cement should necessarily be condemned simply for failure to pass the accelerated tests; nor can a cement be considered entirely satisfactory simply because it has passed these test.

PRINCIPAL PROPERTIES OF AGGREGATES.

General:—Aggregates, being inerts, do not permit of or require as intricate and detailed analysis as cement, which is chemically very active. The requirements are, therefore, chiefly physical, and are for the most part, concerned with the hardness and durability, the size and grading of the various particles. All these things admit of fairly easy and accurate determination, a fact which tends to make the user careless about making any determination at all; nevertheless, they are quite important and, if good results are to be consistently realized, require attention equally as much as cement.

FINE AGGREGATES.

Definition:—Fine aggregates, as usually specified, are such as will pass a screen having no larger than ¼-in. meshes, and sizing from there on down to granules no larger than pin points. Particles failing to pass the ¼-in. screen are then classified as coarse-aggregates.

Character of the Sand:—The sand shall be of hard, dense, tough material. Siliceous quartz sands are the best, although sands ground down from any durable rock will answer Sands formed by the disintegration of soft rock, like slate, shale, or "rotten rock," are not suitable. Sands shall also be free from dirt, slime, vegetable matter, and other evident impurities, and in size be well graded from fine to coarse,—from "pea" to "pin-point". It is very important that they be well graded, for upon the close packing of the grains depends largely the efficiency and economy of the mixture. It is the sand with which the cement comes in most intimate contact, and therefore upon the quality of the sand depends largely the efficiency of the cement.

Grading:—Sands as they come from the bank, pit, or beach are rarely graded properly, requiring either admixture of some coarser grades or separation out of some of the excessively fine material. Occasionally a sand will be found sufficiently well graded naturally, requiring no correction, but such deposits are extremely rare, and even the best deposits will be found to vary considerably as the excavation advances. The user will therefore need to be on the constant look-out and be prepared to make continual readjustment of his proportions.

A very fine sand, or one containing an excessive amount of fine stuff, is highly undesirable. The objections are (1) large percentage of voids in the material, requiring an excessive amount of cement as filler; (2) large surface to coat with cement, the amount increasing as the square of the size of grain, requiring an excessive amount of cement as binder; and (3) a crowding apart of the sand grains by the cement, not being sufficiently larger than the cement grains themselves. The last point is the worst feature. It means that the cement, instead of being pres-

192

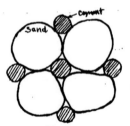

Fig. a
A neat Fit~ Sand Grains
properly sized with regard to
Cement Grains

Fig. b
A bad Fit~Sand
Grains too small, being
Crowded by Cement Grains

FIG. 51—A MAGNIFIED SKETCH SHOWING RELATION BETWEEN
SAND AND CEMENT.

ent in fine points and thin films, filling the final voids and coating
all particles, and thus upon setting locking the entire mass of
particles large and small in a rigid interlacing meshwork, is
bunched and thick-walled. The resulting mortar or concrete is
therefore more fragile and porous, less impermeable, and more
easily acted upon by acids. It is always to be remembered
that cement in body is not a good resister of water pressure or
of acid-attack, and is not tough, so that upon the density
and intimate packing of the inert particles the efficiency of the
concrete will largely depend. It is for this reason, no doubt,
that excessively fine sand has been positively proven unfit
for sea-water concrete. There is always room for a small proportion of fine stuff, even silt, but the bulk of the
sand should be several times larger in size of grain than the cement itself. The bulk of an average Portland cement will pass
a No. 200 sieve, the meshes of which are by design approximately
0.0025 or 1/400-in. The sand should therefore be so sized that
practically all of it will be refused passage by a No. 50 sieve, the
meshes of which are approximately 0.01 or 1/100-in. Its smallest grains will thus be about four times as large as the average
of the cement grains. This relation, it will be seen, will be such
as to allow cement grains to slip into the interstices of the sand
without crowding the sand-grains apart. If, in any sand, the
amount of material passing a No. 60 sieve is slight, not exceeding 10 or 15 per cent of the whole, it may be considered as
equivalent to so much cement as filler, thus effecting a corresponding saving in the amount of cement required without impairing the value of the mixture. This question will be given
separate consideration in a later paragraph.

A very coarse sand, or one containing an excessive amount

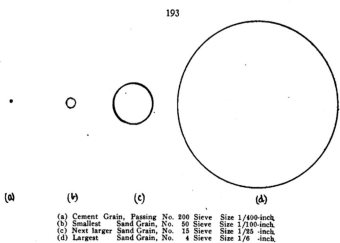

(a)	Cement Grain,	Passing No.	200 Sieve	Size 1/400-inch.
(b)	Smallest Sand Grain,	No. 50 Sieve		Size 1/100-inch.
(c)	Next larger Sand Grain,	No. 15 Sieve		Size 1/25 -inch.
(d)	Largest Sand Grain,	No. 4 Sieve		Size 1/6 -inch.

FIG. 52—COMPARATIVE SIZE OF CEMENT AND DIFFERENT GRADES
OF SAND. ENLARGED APPROXIMATELY 6½ DIAMETERS.

of the maximum size grains, is also undesirable, and for the following reasons: (1) that it requires an excessive amount of cement, is "cement-hungry", so to speak, and (2) that it allows or requires the cement grains to be bunched together, decreasing their relative efficiency and therewith the strength, for reasons as stated in the preceding paragraph. This is the main objection to the *Standard Sand* used in testing, that it contains too large a proportion of coarse grains. Its voids average about 42 per cent, as against 30 to 35 for the best natural sands, and 20 to 25 for an artificially graded sand. Thus it will require an excessive amount of cement and will not show as high results in the strength-tests. By admixing an equal amount of a finer sand, say size about one-fourth, the percentage of voids could be reduced about one-half, a saving of fully 50 per cent in the amount of cement effected, and the strength of the mortar considerably increased.

The question of grading is evidently of prime importance. The appended table, recording tests made by the New York Board of Water Supply, on 1-3 mortars made with sand of different gradings, will serve to enforce the point:

No.	Percentages Passing Sieves.				Tensile Test		Compress. Test		Ratio Compression to Tensile Str.
	No. 4	No. 8	No. 50	No. 100	7 days	90 days	7 days	90 days	at 90 days
1	100	70	12	5	213	613	2690	5640	9.2
2	100	86	21	6	263	412	1915	4660	11.3
3	100	99	26	2	177	325	905	2170	6.7
4	100	97	28	6	178	282	1070	1500	5.3
5	100	94	44	12	139	228	905	1130	5.0
6	100	100	52	14	122	170	275	810	4.7
7	100	100	94	48	80	149	330	490	3.3

The marked superiority of the coarse, well-graded blend is apparent, also the marked inferiority of what would class as excessively fine sands. Significant also is the falling off in the ratio of compressive to tensile strength, which is normal for the first two grades only; or, in other words, while the tensile strength falls off approximately 75 per cent, the compressive strength declines over 90 per cent. Very fine sand is therefore evidently more injurious to the strength of concrete than is apparent even from the tensile tests.

A table of sieves such as are ordinarily employed in the separation of cement and sand is inserted at this point for convenient reference:

No.	Meshes per lin. in.	Meshes per sq. in.	Width of opening ins.	Size of wire ins.	Gauge B and S.
*200	200	40,000	0.00260	0.00240	..
*100	100	10,000	0.00550	0.00450	..
**80	80	6,400	0.0075	0.0050	36
**60	60	3,000	0.0116	0.0050	36
50	50	2,500	0.01440	0.00561	35
40	40	1,600	0.01700	0.00795	32
30	30	900	0.02440	0.00893	31
25	25	625	0.02736	0.01264	28
20	20	400	0.03410	0.01594	26
18	18	324	0.03961	0.01594	26
15	15	225	0.04876	0.01790	25
12	12	144	0.06560	0.01790	25
10	10	100	0.07990	0.02010	24
8	8	64	0.09900	0.02535	22
5	5	25	0.17465	0.02535	22
4	4	16	0.22150	0.02846	21
3	3	9	0.29300	0.04030	18
2	2	4	0.44290	0.05706	15

Note:—The No. in each case indicates the number of openings per lineal inch of sieve. Sieves from No. 80 to No. 4 will be all that are ordinarily required for the analysis of sand.

Density:—The density of sand is a fair measure of its worth, since the best graded sand will invariably have the greatest density,—or in other words the least voids. The fitness of a sand may therefore be partially determined by ascertaining the percentage of voids.

Natural sands will always contain more or less moisture, so that in addition to the ordinary voids, such as the material would have if perfectly dry, therefore known as "air-voids", there will be "moisture-voids". The percentage of voids will also depend upon the compactness. Illustrating how these conditions will vary the determination, the following comparison of sands taken from the same bank, offered by Sanford E. Thompson, is of value:

*Sieves Nos. 100 to 200 ordinarily of brass wire, the diameter of which is approximately equal to the width of opening.
**Sieves Nos. 50 and coarser, of ordinary gauge wire, the diameter of which varies from approximately equal the opening to less than one-eighth the opening. These are conventional divisions, to agree with the wire available.

Voids in Moist and Dry Sand Under Different Conditions of Compacting.

	Percentage of voids.					
	Coarse sand.		Fine sand.		Very fine sand.	
	No. 1.		No. 2.		No. 3.	
Condition of sand	Air voids.	Total voids.	Air voids.	Total voids.	Air voids.	Total voids.
Moist and loose	50	53	56	62	57	63
Moist and shaken	38	42	45	52	46	53
Dry and loose	39	39	44	44	45	45
Dry and shaken	37	37	41	41	41	41

Thus, in this instance, none of the three grades appears especially suited for the purpose upon examination loose and moist, but when dried and compacted all three improve, but No. 1, the coarsest of the lot, makes the most satisfactory showing. It would be considerably further improved by admixture with an equal amount of a finer material, perhaps in this instance of the No. 2 sand.

The determination of voids is very important, not alone as a means of ascertaining the fitness of a sand, but in order to adjust the proportion of cement required. There must evidently be *enough cement to fill all the voids in the fine aggregate*. In the above instance for example, supposing grade No. 1 to be used, the amount of cement required will be at least 37 per cent, or a little over one-third, the volume of the dry, compacted sand. The proportions for a good mortar would thus be 1 part cement to approximately 2½ parts sand, or a 1-2½ mixture, allowing an excess of cement as a compensation for possible variation. If, in this same case, the proportioning had been based upon the void-indication in the condition that the sand was when taken from the bank, a larger proportion of cement would apparently have been required, something say a little better than a 1 to 2. The necessity for drying and compacting a sand before making the determination for voids is evident. Now, if in this case, by an admixture with a suitable amount of a finer grade of material, the actual voids could be reduced to, say 25 per cent, just as strong a mortar could be obtained with proportions as 1 to 4. The saving in cement realizable by improving the grading of a sand is very apparent.

Consistency in Proportioning of Materials:—It should be noted that, if the sand is measured dry and compacted, the cement should be measured likewise. Cement as it falls from the sack or barrel is swelled in bulk considerably. It were better, then, to base the proportions by weight, considering one cubic foot of cement to weigh 117 pounds, which is a fair average for well shaken down cement. The ordinary weight of cement loose is taken at about 100 pounds. Thus, about 1 1/5 cu. ft. of the loose cement should always be allowed for each cubic foot required by a determination based upon measurement of materials compacted. If, on the other hand, the sand as used is moist and loose, the common method of proportioning the mix by measurement of loose volumes will involve small error.

Weight:—The weight of sands is closely related with the question of their density, although it will vary also with the character of the native rock. Quartz sands are thus usually the heaviest, other things being equal. The chief factor in the weight, however, is the proportion of voids. For instance, in the same samples of sand compared in the preceding article, the weights given for the materials under different conditions of moistness and compactness, corresponding to the percentages of voids, were as follows:

Condition of sand.	Weight per cubic foot in pounds.		
	Coarse sand No. 1.	Fine sand No. 2.	Very fine sand No. 3.
Moist and Loose	79	67	65
Moist and Shaken	97	84	82
Dry and Loose	100	92	90
Dry and Shaken	104	98	97

The less, evidently, the percentage of voids, the heavier the sand. Thus, in the heaviest sample in the foregoing table, 104 pounds, the percentage of voids, 37 (confer table in preceeding article), was also the least. The voidless weight of this same sand, that is the weight of a solid cubic foot of it, would be 165 lb. This is the weight of quartz rock, the specific gravity of which is 2.65. This sand then, is ostensibly a quartz sand.

The weight, thus, of a sand, being a measure of the voids, is a good indication of the value of a sand, that sand being in general the most suitable which is heaviest. Natural sands, to make good concrete, as they are, should weigh not less than 80 pounds in the condition that they come from the bank, and when dried and shaken, not less than 100 lb. A 90 to 100 lb. sand, moist and loose, or a 115 to 125 lb. sand, dry and compacted, would be still better.

The weight-property may be made the basis of a table by which, knowing the weight, the voids can be at once read off, thus saving the bother of a separate determination, or conversely, knowing the voids, the weight per cubic foot can be seen. Such a table is herewith appended (page 197.)

Cleanliness:—As to whether dirt in a sand is harmful or not depends upon the character of the dirt. In general this is true—that inorganic soil is, in small percentages at least, comparatively harmless, indeed may be decidedly beneficial, acting as so much fine filler and therefore reducing the amount of cement required, or by helping to close up the pores increasing the imperviousness; while organic soil is as a rule decidedly objectionable, and will require the washing of the sand. Inorganic earth is only so much finely divided mineral matter, like silt or

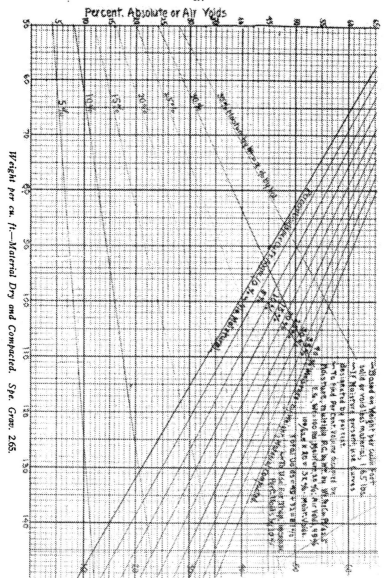

FIG. 53—WEIGHT VOID DIAGRAM.

clay. Organic dirt, on the other hand, is composed of vegetable matter in a state generally of decay. Loam is an example of organic dirt. Loamy sands, therefore, before being used with cement, if used at all, will very generally require careful washing. Some inorganic matter is also detrimental, notably the strong chemicals from manufacturing plants which impregnate the waters of many of our streams and rivers and render their waters unfit to drink and un-inhabitable for fish-life. River-bottom sands are liable to be thus fouled and require, although seemingly clean, careful washing with fresh, clean water. Salt-sea-bottom sands are also open to objection for a similar reason, the inevitable salt in this case probably being magnesium sulphate which reacts injuriously with the cement, causing it to swell and blow. All these things need to be watched very carefully, for very many times failures of cement construction scored against the cement are in reality due to the quality of the sand. Slimy sand is bad, for the ooze prevents the cement from coating the particles Clay in sticky lumps is also bad, for these lumps ball up the mixture, interfere with the coating of the particle with cement, and produce soft or "rotten" spots in the concrete. Clay, if present, should be in a very finely divided condition. Mica is another objectionable element, especially for surfacing material, as the little flakes float up to the top and cause the surface to dust and peal. The color of a sand is an indication, although by no means a conclusive one, of a sand's cleanliness. Beach sands are nearly always pure white, but bank sands perhaps every bit as good material for concrete, will appear a dark or reddish-brown. The dark color may be due to the presence of harmless iron salts. If the color is due to harmful dirt, it may be simply detected by rubbing a bit of the material in question against the palm of the hand. A dirty stain, probably odorous, indicates undesirable matter. Or, it may be shown by stirring a handful of the material in a glass of clean water—muddy discoloration being the sign. A yellowish turbulence however, may only indicate clay.

That clay in sand, even in considerable percentages, is not especially detrimental to the strength, the following table of test data would seem to indicate:

| | | Average tensile strength pounds per square inch. | | |
Material	Per cent of clay in sand.	Per cent of water used in mixing.	1 day in air, 6 days in water.	1 day in air, 27 days in water.
A—Neat Portland Cement	....	21.0	527	862
B—Standard Quartz Sand (1 part cement, to 3 parts sand)	....	9.9	175	263
C—Gravel (unwashed, unscreened) (1 part cement to 3 parts gravel)....................	25.0	11.2	208	316

D—Gravel (washed but unscreened) (1 cement 3 gravel)	3.0	10.7	230	355
E—Gravel (scoreened but unwashed) (1 cement 3 gravel)	25.2	13.7	219	335
F—Gravel (washed and screened) 1 : 3....	3.2	11.2	211	367
G—Trap Rock Dust. (1 cement to 3 grit) 20 per cent retained on No. 20 sieve...........	16.4	15.0	213	385
H—Trap Rock Grit (1 : 3)...................	11.4	12.5	282	459
J—Crushed Trap Rock (1 : 3)...............	....	11.2	279	416
K—Trap Rock Grit and Excav. Gravel, 1:1½ Grit; 1½ Gravel (unwashed, screened).....	....	12.5	218	372

*Compiled from tests made by the Philadelphia Rapid Transit Commission, and reported by Mr. Charles M. Mills before the Philadelphia Club of Engineers, July, 1904.

The best results are seen to have been given by the crushed rock mixtures, the poorest by the Standard Quartz Sand. The effect of the clay is, as far as the strength is concerned, practically negligible.

The following diagram, due to Mr. Philip L. Wormley, Jr., Testing Engineer, Office of Public Roads, U. S. Department of Agriculture, is also of value as indicating the effect of clays upon sand:

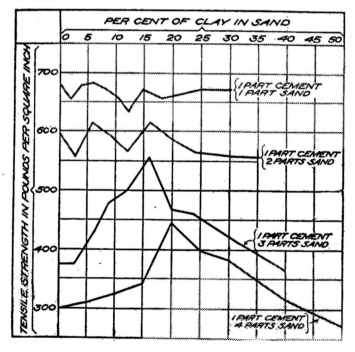

FIG. 54—DIAGRAM SHOWING EFFECT OF CLAYS ON SAND.

The sand used was a crushed quartz, having approximately 43 per cent voids, and a clay similar to that commonly found in sands. All proportions were by weight, and tests made at 90 days. The clay in this case, in percentage up to 15 per cent and even higher, is seen to be markedly beneficial.

Illustrating that the presence even of a small percentage of loam may be harmless, indeed on short time tests somewhat beneficial, the following diagram platted from tests reported in the *Engineering News,* April 28, 1904, by Mr. G. I. Griesenauer, is of interest:

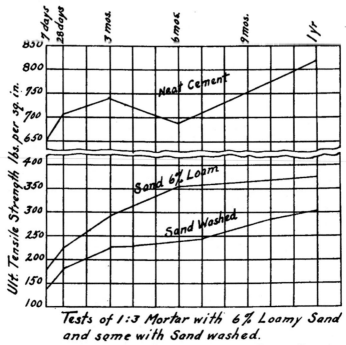

Tests of 1:3 Mortar with 6% Loamy Sand and some with Sand washed.

FIG. 55—COMPARATIVE TESTS ON CLEAN AND LOAMY SAND.

The difference between *clay* and *loam* should be noted: Both are finely divided materials, but the clay is powder ground off rocks and boulders and is thus naturally affinitive to both the sand and the cement, while the loam is a mixture of vegetable mold. Clay up to 25% of cement may not impair the strength, rather actually enhance it by improving the grading—increasing also the impermeability and effecting a saving in cement. Loam, on the contrary, even in amounts as small as 5 per cent, may prove detrimental. One of the advantages of clay in this connection is its colloidal property, by which is meant its ability to puddle or merge, becoming impervious to moisture. It is this property of clay that makes it useful in earthen dam construction, in which in the form of a core wall it forms an effectual barrier to the passage of water. So likewise, its presence in concrete mixtures, in proper amounts properly distributed, contributes to the water-tigntness of the concrete, and becomes thus on occasion a valuable adjunct.

There is, however, another consideration, independent of the question of strength, that may decide the question,—namely, the surface finish required. Dirt in the sand, which may not impair the strength in the slightest, yet may produce unsightly discoloration of the surface, and therefore be unsuited for surface work. Sand for fine surfacings will in general require careful washing, in order to insure freedom from discoloration and uniformity in shade.

The benefit of clay in sand may be said in general to vary directly with the coarseness and amount of voids, that sand being most benefited that is on the whole the coarsest and contains the most voids. Contrarywise, a fine sand or a well graded sand would fail to profit, instead very likely be harmful. No general conclusion may therefore be stated as to the desirability of clay, its presence or its absence being a matter for separate determination in each case.

Clay should scarcely be added deliberately to a concrete mixture for the sake of increasing its impermeability,—instead, slaked lime or puzzolan, which in proper form answer the same purpose and in a better way; being themselves cementitious, they contribute to the strength of the concrete. Clay should be regarded in general as a "necessary evil," to be tolerated on occasion but not condoned, certainly not licensed.

The washing of a sand of course adds to the cost, say from 10 to 25 cents a cu. yd., depending on whether it is washed by machine or by hand and the quantity handled; consequently it should always be predetermined as to whether the dirt present is actually baneful before undertaking to wash.

Color:—The color of a sand is important chiefly as it concerns the outward appearance. There are sands of all colors and shades, from the pure white beach sand to the blackish river sand and the reddish-brown bank sand, and the desirability from a strength viewpoint may be, and generally is, in inverse to the order of desirability as regards appearance. For most above ground work, especially exteriors, the lighter the sand the better, and for fine architectural work a pure white sand is desirable. The selection on this score is thus one very largely of taste.

The color of the sand makes less difference to the color of

the finished concrete than might be supposed. This is due to the bleaching action of the cement.

The following table indicates in general the effect of different colored mineral pigments on cement mortar. The precise shade obtained in any case will of course depend also on the color of the cement and the color of the sand. If a pure white mortar or facing is desired both a white sand and a white Portland cement should be employed, although the white Portland is unessential if the surface is to be tooled or acid-etched.

Colored Mortars.

Colors Given to Portland Cement Mortars Containing 2 Parts River Sand to 1 Part Portland Cement.

Dry Material Used.	Weight of coloring matter per 100 lb. of cement.				Cost of coloring matter per lb.
	½ lb.	1 lb.	2 lb.	4 lb.	
*Lampblack	Light Shade	Light Gray	Blue Gray	Dark Blue	15 cts.
Prussian Blue	Light Green Slate	Light Blue Slate	Blue Slate	Bright Blue Slate	50 cts.
Ultramarine Blue	* *	Light Blue Slate	Blue Slate	Bright Blue Slate	20 cts.
Yellow Ochre	Light Green			Light Buff	3 cts.
Burnt Umber	Light Pinkish Slate	Pinkish Slate	Dull Lavender	Chocolate	10 cts.
Venetian Red	Slate, Pink Tinge	Bright Pink Slate	Light Dull Pink	Dull Pink	21½ cts.
Red Iron Ore	Pinkish Slate	Dull Pink	Terra Cotta	Light Brick Red	21½ cts.

Coloring matter should be used with caution, as its effect on the cement has not yet been very definitely determined. Certain pigments are known to be injurious; for instance, red lead, the pigment of the paints used for coating iron work, is even in minute percentages decidedly bad for concrete. If used at all they should be in finely powdered form and intimately intermingled *dry* with the cement. This matter interests the maker of architectural effects more than the average worker in cement, and the concrete man hardly at all. It is not good taste to tinker with the truly delightful natural soft gray tints of concrete work, and the chief concern on this score will be to see that the brand of cement and the grade of sand used shall be consistent throughout a single piece of work, to the end that unsightly variations in shade and tone be avoided.

As to whether coloring matter present in a sand is harmful will of course depend largely on what it is: If of mineral nature or *insoluble*, it will in general be harmless; if vegetable or *soluble*, the reverse. The determination of *cleanliness* will usually decide this point as well. In any case, what causes the color should be ascertained, the sand tried with cement to determine its effect on the setting qualities and tensile strength, and the treatment decided accordingly.

*Common lampblack (not of bone), as well as Venetian-red, are worthless as they will run.

Artificial Sands:—It has become the practice of late, in event of a suitable natural sand being unavailable, or requiring an inconvenient and expensive haul, to manufacture the sand to order. The operation is in two stages: (1) To crush the rock as ordinarily, then (2) to screen out the finer particles and reduce further by running through rolls. The resulting material will compare very favorably in grading with the best natural sands, containing, however, an excess of very fine material or dust. Such sands will often, if not in general, show higher strength than natural sands. An instance of this has been seen in the tests recorded in this chapter. Usually it will be found, however, that the best results will follow the blending of an artificial sand with some natural sand, the same being both stronger and more impervious than either one alone.

An advantage of artificial sand is that the stone for it can be carefully selected, and that when crushed it will contain no injurious matter. The prime disadvantage is the liability to excessive dust, which would open the material to the same objections as to excessively fine sand. Very often, too, the dust will be segregated in pockets, which is of course highly objectionable.

Trap, basalt, quartz and hard limestone make the best artificial sand, although good sandstones, granites and conglomerates make satisfactory material. A sand is also manufactured out of slag. Trap sand will make the hardest and most endurable, limestone the strongest and least permeable—but less durable mortar or concrete.

In general, what applies to natural sands applies with equal force to artificial sands.

Use of "Crusher-Run:"—The availability of sand of this sort means that broken stone aggregates may not require screening, as commonly specified, but may be used as "run-of-crusher" without danger. In this case the proportion of sand required in addition may be little or nothing, a fact which should always be borne in mind. The writer has seen crusher-run used as if it were screened stone and the stated proportion of sand added religiously, with the result that the mixture was exceedingly "mushy," and afterwards, in spots, crumbly. As, owing to the unavoidable tendency of grades to separate in the several handlings intervening between crushing and using, the fine material is more than likely to be bunched, it will in general be necessary to screen out the fine stuff and recombine as used.

Laws of Grading:—It has been seen that the principle of grading is that the finer grade should slip nicely into the interstices or voids of the coarser. There are evidently two governing considerations in the grading, first, that the different grades be properly sized, and, secondly, that there be a proper amount of each size. If any interemdiate grade is sized too large it will crowd the next larger size and bunch the next smaller; or, if there is a deficiency of any one size the result will be a crowding of all sizes above. The effect, in any event, is a mixture less dense, more pervious, less strong, and requiring more of the cement—the costly element—to produce a given result. Now, if the grains were all perfectly spherical, it would be possible to calculate the exact size and amount of each grade to produce the maximum reduction of voids. But close examination of the particles of a sand will disclose all manner of shapes, from rounded to angular. Moreover, an absolute scientific blending would probably require the elimination of many of the intermediate sizes of material, which would be impractical to do. Therefore, we must rely mostly on experiment and experimental data to guide us

Let us pause to consider what loose volumes of material actually means. The voids, or air spaces, in a granular material are said to be so many per cent, implying that' if the material were solid there would be no voids. But this is a literal impossibility. No matter how dense a solid block of material may be, it would always contain some voids, which, although indeterminate by ordinary methods, yet would be plainly visible under a powerful magnifying glass. This must forever be the case, from the molecular structure of all material substance. There will always be interstices between the points of contact of the infinitesimal particles, themselves globular. But the finer the division of particles in any mass is carried out the denser it will be, that is, the greater the reduction of voids. Now let us reverse the process. Suppose a solid block of material, say. stone, measuring one cubic foot, with voids virtually *nil*,

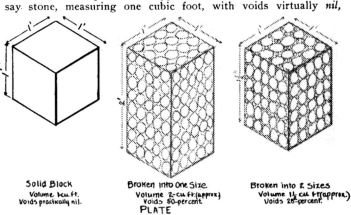

Solid Block
Volume 1 cu. ft.
Voids practically nil.

Broken into One Size.
Volume 2-cu.ft.(approx.)
Voids 50-percent.
PLATE

Broken into 2 Sizes
Volume 1½ cu ft.(approx.)
Voids 26-percent.

FIG. 56—SHOWING EFFECT ON VOLUME AND DENSITY OF
CRUSHING A SOLID BLOCK OF STONE.

to be broken into a large number of equal sized pieces. The volume would be found to increase to approximately double or 2 cu. ft. This is generally true, that all granular material of one-dimension grains has a density roughly of only one-half, or voids in the neighborhood of 50 per cent. Now suppose a portion of the first-sized pieces, say about one-quarter of them, be broken up into smaller fragments, say of about one-quarter the diameter of the first size, and that the two grades be thoroughly intermingled. The net volume will be reduced to approximately one-and-a-half cubic feet, and the voids to approximately 25 per cent. And so, by breaking a portion into still smaller size, a further reduction of volume and voids could be effected; ad infinitum, approaching the original volume of 1 cu. ft. as a limit. The volume at any stage in the reduction will be very closely that of the largest size pieces. This is the principle of concrete proportioning, and explains the phenomenon, sometimes puzzling to workmen and untutored workers, whereby 25 cu. ft. of cement, sand, stone and water only produce 16 or 17 cu. ft. of concrete, which is but little more than the volume of stone or large aggregate used. If the material were exactly sized and graded, the final volume would be that of the large aggregate, and the more closely this condition is realized in practice the denser and stronger the concrete and the less the amount of cement required to produce the result. In this case, as usually, efficiency and economy go hand in hand.

The *chief points of interest* in connection with an aggregate are thus *the size and amount of each grade and the limiting size either way,* that is, *the minimum size* and the *maximum size.* Our starting point is the cement—as finest filler. The desideratum is that the cement grains be distributed as widely as possible, or, which is the same thing, that they are not caused to segregate or bunch. We have seen that the fineness of the average Portland cement is such that the bulk of it will pass a No. 200 sieve, the meshes of which are 1/400 inch (0.0026). The smallest size of sand grain, therefore, should be approximately (as may be demonstrated graphically) three to four times the size of the cement particle, or such as will pass about a No. 60 sieve, the meshes of which are 0.0116 inch, and be refused by a No. 80, the meshes of which are 0.0075 inch, and the amount of this grade should be approximately twice that of the cement (by volume). Similarly, the next larger size should be approximately four times as large as the preceding, or 16 times as large as the cement grain, and there should be about twice as much of it as the cement and finest sand put together, the cement being considered as already absorbed into the volume of the finest sand. And so forth, in like ratio as to size and quantity, up to the maximum size of aggregate that may be considered to combine in this manner. For the average sand, the maximum will be such as will pass just a No. 4 sieve, the opening of which is 0.2215 inch (practically one-quarter-inch). It needs be remarked that in all these determinations, it is convenient to use weights rather than volumes, although volumes are concerned. This will involve no error, since equal volumes of evenly sized grains of different size of the same kind of stone will be identical in weight. To express these relations in percentages: All, or 100 per cent, should pass the No. 4 sieve; approximately 50 per cent by weight of the whole the No. 15 sieve; about 25 per cent the No. 60 sieve; and all that then remains refused by the No. 80 sieve. If a small percentage—5 to 10 per cent—passes the No. 80 sieve, it need not be rejected, but considered as equivalent to so much cement as finest filler. If a larger percentage than this, then some of it should be removed if the maximum strength concrete is desired. There will, of course, be varying amounts of intermediate sized grains, which will tend to some extent to mar the perfection of the blending, but practically it would be next to impossible to do any better.

To ascertain the grading of the intermediate sizes—that there is not an excess or a deficiency of any intermediate grade— a number of sieves should be interposed between the main sizes. Otherwise, by the mere fact of passing one of the determining sieves, the percentage of that size material may not be established. For example, 50 per cent may be found to pass the No. 15 sieve and all be rejected by the No. 60, but all but 10 per cent pass a No. 40 sieve. This would indicate a deficiency in particles as large as No. 15, and an excess of particles between No. 40 and 60. The more complete the sieving, that is the greater the number of sieves used, the more accurate the determination. All results should be carefully tabulated in a record book for use as indicated in the next paragraph.

Curves of grading:—The above relations may also be expressed diagramatically. On the horizontal the sieve openings are laid off on an exaggerated scale in ten-thousandths-inch, and the number of sieve corresponding marked down. On the vertical, the percentages, from 0 to 100. The curve is drawn at the proper intersections. It intersects the 100 per cent line at the No. 4 vertical, which means that all of the material passes the No. 4 sieve. So at the No. 15—50 per cent intersection, which the curve cuts, the signification is that 50 per cent of the material passes the No. 15 sieve. Similarly for any intersection. There have been platted for comparison the analyses of the sands experimented upon by the New York Board of Water Supply, as recorded formerly in this chapter. The curves marked No. 1 and No. 2 gave the best results in the one to three mortar tests. It will be seen

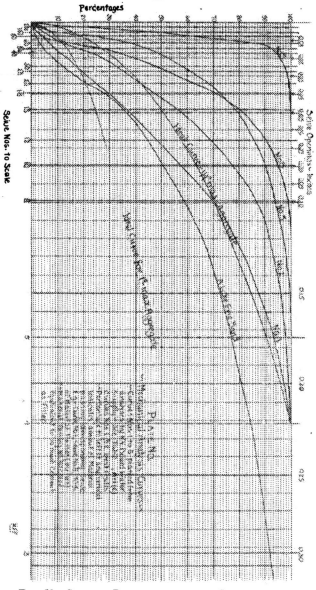

FIG. 58—GRAPHIC REPRESENTATION OF SAND ANALYSIS.

that the author's curve lies approximately between these two, slightly exceeding both for the finer sizes, which indicates that more of the finer grades could have been used in both with benefit. The agreement between the three curves, No. 1 and No. 2, and the author's, is remarkably close, and it will, therefore, be safe to formulate the Law of Grading upon this basis. There has also been platted an analysis of a Lake Erie sand, in which a larger size has been made the basis. It will be noticed that it very closely parallels the empirical curve. The sagging in the lower end indicates a deficiency of fine material. Otherwise the indication of blend is excellent. This would be a good gravel analysis curve. By platting the results of an analysis of any sand on this diagram its value would be at once apparent by comparison with the standard curve. Thus, of any number of sands compared, that which closest approximated the standard would be the one to select.

Some of the conclusions stated by Messrs. W. B. Fuller and S. E. Thompson, members American Society Civil Engineers, and published in the *Transactions* of that society for December, 1907—Vol. LIX, page 70, are of interest in this connection as affording further substantiation of the foregoing:

1.—"The best mixture of cement and aggregate has a mechanical analysis curve resembling a parabola, which is a combination of a curve approaching an ellipse for the fine aggregate portion and a tangent straight line for the coarse aggregate portion. The ellipse runs to a diameter of one-tenth the diameter of the maximum size of aggregate, and the aggregate from this point is uniformly graded.

2.—"The ideal mechanical analysis curve, i. e., the best curve, is slightly different for different materials. Cow Bay sand and gravel, for instance, pack closer than Jerome Park stone and screenings and therefore require less of the finer gradings.

3.—"The form of the best analysis curve for any given material is nearly the same for all sizes of aggregate, that is, the curve for ½-inch, 1-inch and 2¼-inch maximum aggregate may be described by an equation with the maximum diameter as the only variable. In other words, suppose a diagram in which the left ordinate is zero, and the extreme right ordinate corresponds to 2½-in. stone, with the best curve for this stone drawn upon it. If now, on this diagram the vertical scale remains the same, but the horizontal scale is increased two and one-quarter times, so that the diameter of the 1-in. stone corresponds to the extreme right hand ordinate, the best curve for the 1-in. stone will be very nearly the one already drawn for the 2½-in. stone. The chief difference between the two is that the larger size of stone requires a slightly higher curve in the fine sand portion.

4.—"It follows from the last conclusion that, from a scientific standpoint, the term 'sand' is a relative one. * * *

5.—"The strength and density of concrete is affected by the variation in the diameter of the particles of sand more than by the variations in the diameter of the stone particles.

6.—"An excess of fine or medium sand decreases the density and also the strength of the concrete, as will also a deficiency of fine grains of sand in a lean concrete.

7.—"The *substitution* of *cement* for *fine sand does not affect* the *density* of the mixture but *does increase* the *strength,* although in a slightly smaller ratio than the increase in the ratio of cement.

8.—"It follows from the foregoing conclusions that the *correct proportioning* of *concrete* for *strength* consists in finding, with *any percentage* of *cement,* a *concrete mixture* of *maximum density,* and *increasing* or *decreasing* the *cement* by *substituting it* for the *finer particles* of *sand* or *vice versa."*

Practical Application of Curves of Grading:—The practical application of the sand-analysis diagram is apparent. Curves* 3, 5 and 7, which swell considerably above the best curve, contain an excessive amount of very fine material, especially No. 7, and this, too, was seen to give the poorest results in the tests. Curve No. 1 is slightly deficient in fine material, which requires

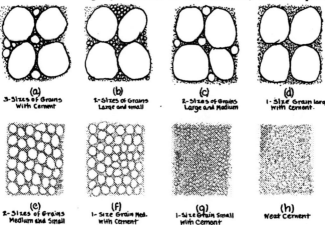

(a)
3-Sizes of Grains
With Cement

(b)
2-Sizes of Grains
Large and small

(c)
2-Sizes of Grains
Large and Medium

(d)
1-Size Grain large
with cement.

(e)
2-Sizes of Grains
Medium and Small

(f)
1-Size Grain Med.
with Cement

(g)
1-Size Grain Small
with Cement

(h)
Neat Cement

Fig. 59—Showing Effect of Various Possible Combinations of Sized Grains of Aggregate and Cement. (a) is best graded, densest, strongest, requiring least cement. Efficiency decreases regularly, (g) being least efficient while requiring most cement. (h) represents mass of neat cement, which while strong is ineffective because shrinkage is too great. Relative shrinkage is maximum for (h), decreasing regularly to (a), which has minimum.

*Fig. 58, Page 207.

compensation with an equivalent in cement. Suppose a curve of an analysis sagged very badly from the maximum size down, the indication would be that there was a marked deficiency of all finer grades, requiring a large percentage of cement. Such a case is represented crudely in sketch (d), Fig. 59. In this case 50 per cent as much cement as sand will be required to fill the voids. The grading represented in sketch (a) is the ideal and its curve would be found to closely approximate the standard. The various possible conditions are represented on the other sketches on this plate. The curve for (g), for instance, would rise rapidly to the 100 per cent—No. 60 sieve intersection, which is a condition very similar to curve No. 7 already drawn. The reason for the extremely poor showing of this particular sand is now apparent. A 1 to 3 mixture was used. From the sketch it is evident that this proportion leaves fully one-half of the voids unfilled, a sufficient cause for weakness. For a sand of this character no leaner than a 1 to 2 should have been used. As compared with this the mixture represented by sketch (a) and the standard curve will require, for a perfect filling of voids, something like 15 per cent, which means that a mixture as lean as 1 to 6 or 7 would, with it, be as good as a 1 to 2 with the other. Indeed, it would, in many respects, be far better, being more dense, more impermeable, and considerably tougher. *The very great saving in cement, coincident with the all around improvement of the mixture, is manifest, and is a fact of paramount importance.*

Maximum Size Fine-Aggregates:—This same method of analysis may be extended to embrace sizes of material as large as will enter freely into the fluid-consistency of the mix and not separate in handling, the curve of analysis grading in a parallel manner from zero to an intersection with the 100 per cent line at the diameter of the maximum grade. However, in general, it will be more convenient to consider that the blending proper ends with the coarsest sand (or fine gravel) and that sizes above this act simply as "bulk-swellers", or, in other words, that the mortar acts merely to fill the voids in the large aggregate. In this event, then, the *minimum size "bulk-swelling" aggregate* should not be less than 4 times the diameter of the maximum size of fine aggregate—in the case of the typical sand analysis considered in this article, 4 times 0.2215 in., or between ¾ and 1 in. Any size above 1 in. will be equally as satisfactory, the larger possibly the better. Messrs. Fuller and Thompson, in the same paper as aforenoted, offered also the following pertinent conclusions:

(1) "Stone of the largest size makes the strongest concrete under both compression and transverse loading, i. e. a graded aggregate in which the maximum size stone is 2¼ in. in diameter gives a stronger concrete than a graded aggregate with 1-in. maximum size, and the 1-in. stone in turn gives a stronger concrete

than a ½-in. stone. A concrete in which the graded aggregate runs to 1 in. in maximum size will require for equal strength about 1/6 more cement, and with an aggregate running to ½ in. maximum stone about 1/3 more cement, than concrete with an aggregate in which the maximum size is 2¼ in.

(3) "The largest stone makes the densest concrete. Concrete made with graded stone having a maximum diameter of 2¼ in. is noticeably denser than with 1-in. stone, and this in turn than with ½-in. stone.

(4) "The strength and density of concrete is affected but slightly, if at all, by decreasing the quantity of the medium size stone and increasing the quantity of the coarsest stone. *An excess of the medium size, on the other hand, appreciably 'decreases the density and strength of the concrete."*

Proper Sizing of Coarse Aggregate:—The importance of having the coarse aggregate sized properly with relation to the coarsest fine aggregate is very evident. The medium stone instanced in (4) are, no doubt, too large to incorporate with the mortar or to slip in freely into the voids of the coarser stone, thus crowding the particles of the latter apart and interfering with the proper functioning of the former. Elimination altogether of this size, then, would be an actual benefit to the mixture.

Experiments indicate that as much as the coarse, "bulk-swelling", aggregate may be used, indeed should be used, as, mixed with the graded mortar, will produce a concrete in which the voids are just nicely filled. This will require, assuming the voids in broken stone of one size at approximately 50 per cent (or permit of, as one chooses to look at it) two parts of the coarse stone to one part of the graded mortar. These proportions shall be of course so varied as to produce a mixture in which there are no visible voids and all stone covered.

Mr. W. B. Fuller, again, has furnished data that substantiate the point at issue. The table speaks for itself.

Proportions by Wt. of Cement to Total Aggregate	Proportions by Wt. of Cement to Sand and to Broken Stone.	Breaking Stress, Outer Fiber, lbs. per sq. inch.
1-6	1-1-5	504
1-6	1-2-4	439
1-6	1-3-3	355
1-6	1-4-2	210
1-6	1-6-0	93

Tests made on 6-inch beams, at Little Falls, N. J., and reported in Taylor and Thompson—"Concrete: Plain and Reinforced."

Summarizing, the Laws of Grading May be Stated as Follows:— 1. The size and proportions of the grains in the fine-aggregate shall be such as, with the due amount of cement and water, will produce a mortar of maximum density,—a condition that may be represented by a curve of analysis, the horizontal distances of which are the sieve openings to scale, and the verticals the percentages by weight passing each size sieve. The best curve will in general be such as will intersect the 100 per cent line at the No. 4 sieve vertical, the 50 per cent line at the No. 15, and the 25 per cent at the No. 60, shading from there on to zero. Not more than 10 per cent should pass the No. 80 sieve, since from that point the cement comes into consideration. The small amount passing the No. 80 may be considered to replace so much cement as finest filler.

2. The *minimum size of coarse-aggregate* shall not be less than will allow the graded mortar to slip in freely between the particles without crowding them apart, and the *proportion* such as, with the mortar, will give a concrete smooth and dense appearing, with no visible voids or stone particles uncovered.

The specific size will depend upon the maximum size of fine-aggregate. If the latter is approximately ¼-in., the minimum size coarse-aggregate shall not be less than four times as large, or 1-in., and preferably larger if the conditions of usage permit. If the limiting size (maximum) of coarse-aggregate is only ½-in., then the fine-aggregate shall not be sized larger than ⅛-in., or such as will all pass a No. 8 sieve, and preferably still finer, such, say, as will all pass a No. 15 sieve. The coarser blend all around is much better, and should be used whenever the occasion permits.

3. If only one class of aggregate is used, such as a gravel or "run-of-crusher", it may be analyzed in a like manner, all distinction between "fines" and "coarses" being in this event merged. The mechanical analysis curve in this case would be based upon the maximum grade of gravel, conforming in general character with the Ideal Curve constructed for the No. 4 sand, but with its per cent points correspondingly advanced and lowered.

4. The *proportion of cement* shall be such as will generously fill the voids in the graded aggregate, allowing at least 10 per cent excess over the net indication as a compensation for probable variations and discrepancies in the materials, and, in the case of reinforced-concrete, to insure also an excess of fine material for bonding purposes and to protect the steel. It may even be necessary on occasion—and it is a wise measure in any event—to slush a little special rich and fine mixture in and around the steel as the deposition of concrete takes place.

Assuming 100 lb. cement to a cubic foot, 1 lb. of cement will evidently be required for each per cent, voids per cubic foot of aggregate—*combined* and *compacted*. Allowing 10 per cent excess, 1.1 lb. will be required. A diagram is herewith presented based upon this relation, by use of which—knowing the cubic foot weight in any case whatsoever of combined and compacted aggregate—the proportion of cement in pounds per cubic foot of material can be looked out. Curves are also provided for taking into account the proportion of clay or excessively fine insoluble material it may be desired to substitute for a portion of the cement. But obviously, if such be already present in the aggregate when the weight per cubic foot is found, no correction need be made. The percentage of cement by weight, however, should be less than 10 per cent; this is the safe minimum.

It may, on occasion, be beneficial to replace part of the cement required with *lime* or *puzzolan*, especially if water-tightness is an essential. The action of the lime (as hydrate, Ca (OH)$_2$), apparently is to increase the formation of colloids. Recent experiments indicate, that up to 25 per cent of the cement can with benefit be replaced by lime,—not only decreasing the permeability but increasing the strength and economy. Lime has long been a legitimate co-partner in cement mortars, improving the quality in amounts up to as much as 50 per cent. Sabin, in his publication *"Cement and Concrete"* (1905), reports finding 50 per cent lime to more than double the cohesiveness or strength of the mortar, and 25 per cent to increase it over 150 per cent. Besides, lime even in less percentages, markedly improves the workability of the mixture,—causing it to "butter" more freely, as the mason terms it. The lime, need it be explained, should be hydrated, that is slaked, and after hydration, dried to remove all surplus moisture and powdered; in such form only can it be freely and sufficiently intermingled with the cement. Such a product is marketed especially for this purpose. The lime should be practically pure calcium hydrate, with little or no magnesia in combination, limes high in magnesia not being suitable or safe for use in conjunction with Portland cement. The advisability of introducing much lime into cement-concretes may be questioned; however, concrete being itself a form of mortar, the same reasons that justify its use in ordinary mortars vouch for it here. There is an additional advantage, in case the concrete is reinforced, due to the increased alkalinity of the mixture, in that the imbedded metal is thereby further safeguarded against corrosion, steel in a strong lime solution being immune to rust.

Puzzolan, added to Portland cement mixtures, acts similarly to lime, increasing the formation of colloids and therefore the impermeability and strength. *Trass* is the trade name for the puzzolan in commonest use for this purpose. Blast-furnace slag, being akin to puzzolan in composition, may also be admixed; but, likely to carry an injurious amount of sulphides, it is not so safe. This, it will be recalled, is the objection to slag cements. (See Chapter I.)

Sometimes *both lime* and *puzzolan* are admixed, substituting each about a third of the cement; it is doubtful if it pays to complex matters to this extent, however. Again there are those who endeavor to dispense with the cement altogether, using simply the trass and lime, which is nothing more or less than a reversion to the primitive puzzolan cement, such as perhaps the Romans used. William Challoner, who seems to favor such a combination for some purposes, has on record* this: "Trass-lime mortars are superior ultimately in strength and much cheaper than the non-dense or weak cement-sand mortars. . . After seven days a mortar of 1½ trass, 1 lime and 1 sand becomes stronger than 1 of cement and 3 of sand, and nearly attains the resistance of 1 of cement to 2½ of sand."

While all of the above combinations may have favor, and on occasion their legitimate applications, it is doubtful if they should be used in high class, especially reinforced concrete, construction. The virtue of Portland cement concrete has been too well proved to indulge in questionable experimentation, and while a certain amount of admixture with lime is no doubt a benefit, particularly if impermeability is a desideratum, in general it will be *wise to hold fast to the importance of always having a sufficient amount of good Portland cement used in the right way.* If a portion is to be replaced with lime, the percentage should not exceed 25 per cent, and better not 15.

Correction of a Sand:—In event of the impracticability of selecting a sand along the lines of the foregoing, no doubt because none of the available sands are found on analysis suitable, the best sand available may be corrected by addition of suitable materials.

The method of procedure is simple. A sample of the sand should first be separated into its grades, its curve platted and the deficiences noted. Then an endeavor should be made to secure a supply of a suitable grade of material, either sand or screenings, to remedy the deficiency. If a suitable grade is not available, or the size of the job does not warrant the expense of importing it, the deficiency will have to be supplied with cement, and, thus, the best made of an unfavorable situation. The proportion of cement required may be found on the same diagram, since the deficiency will show in a reduced weight per cubic foot of the combined aggregate.

As an illustration, suppose the analysis is such as indicated by Curve No. 7, page 207, of which 95 per cent. passes the No. 50 sieve, and 100 per cent the No. 15. Then, to correct the blend, there will be required approximately again as much of a coarse sand grading between sieves No. 15 and No. 4. The corrected sand will then be found to give a curve closely approximating the ideal, since it is now constituted so that only 50 per cent will pass the No. 15 sieve. And similarly in any case. Suppose again, that the material were mostly maximum sized, being such as would all pass the No. 4 sieve but he caught on the No.

*In the "Engineer," Sept. 6, 1907, pub. in London.

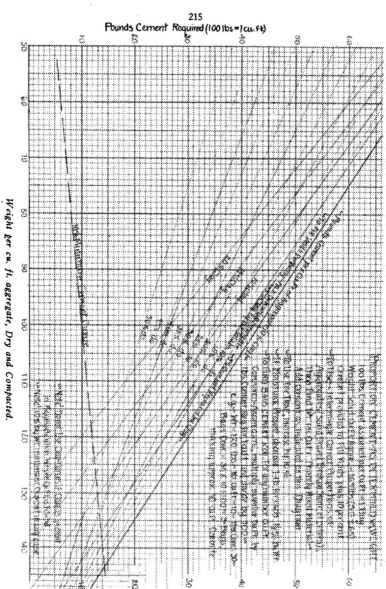

FIG. 60—PLATE SHOWING PROPORTION OF CEMENT AS DETERMINED
BY WEIGHT.

15. Then, half as much, or 50 per cent of a finer material, grading from No. 15 down, will be required. In some cases the admixture of three or more grades of material may be required. Leonard B. Wason, president of the Alberthaw Construction Co., Boston, relates, on one job, of sizing his aggregate (a gravel) into five (5) separate bins, and recombining in proportions as determined by tests, effecting thereby a saving in cement alone that more than counterbalanced the expense and trouble of testing and sizing.*

The method of procedure on work will be to size the material as it is received into separate bins, recombining in the indicated proportions as used. The precise manner of accomplishing this separation and recombination is a detail of practice, and simple and economical ways of doing it will readily suggest themselves on occasion to the practical constructor.

Complete Mechanical Analysis:—A complete mechanical analysis of aggregates should thus include the following steps:

1. Make a first selection of material on the basis of weights per cubic foot when dried and compacted. The heaviest sand has been seen to be the one with the least voids. Check this choice as follows:

2. Mix each material proposed for choice into a mortar, of a plastic consistency, basing the amount of cement used upon the void indication of the weight, or if proportions are specified according to the stated proportions. That grade will be most suitable which produces a mortar of least volume, and which at the same time is the smoothest appearing, with no visible voids. The heaviest aggregate should also give the least volume. This is the *Volumetric Determination,* and may be made as follows:

Volumetric Determination:—Required a good sized bucket. A 16 quart pail such is found on every concrete job will answer. The larger the vessel the more reliable the indication. Measure out equal volumes of each of the materials to be compared, such, say, as will fill the bucket about three-quarters full when well shaken. Record the height of material in the bucket by a notch on a stick of wood, which should be at least long enough to extend through the material to contact with the bottom with several inches to spare. Then mix each in turn into a mortar of soft, plastic consistency, but not so soft as to run, using a proportion of cement as indicated by the weight-void table, and water by judgment. Pour each mixture in turn into the vessel, compact by shaking and kneading with the stick, and record the level it takes by another notch. That material will be best which rises the least above the original notch, and at the same time makes the smoothest appearing mortar, indicating proper filling of voids.

*Concrete Engineering, May, 1909, Page 135.

If, as is very often the case, it is not feasible to make a large scale determination, the operation may be conducted in a glass or a bowl. It is advisable, however, in this as in all testing, to work with as large volumes as practicable, since the larger scaled the test the more valuable the indications practically.

3. Secondly, having made a choice of material,—or starting with any material at hand,—to correct it, make a separation of it into several grades, plat its curve and compare it with the best curve for that grade of material. The deficiencies will then be readily apparent and determinable. Rectify the deficiency or deficiencies by addition of a suitable amount of the proper size or sizes, and thoroughly blend.

4. To verify correction, weigh equal amounts by volume of the original material and the corrected blend. A suitable increase in weight indicates that the determination is satisfactory. To secure accurate results, both materials should be first dried, to remove all moisture voids, and then thoroughly compacted by shaking, and the "doctored" blend in particular well shaken to insure that it is in truth *blended*. Drying may be conveniently accomplished by spreading out the material in a thin layer on some flat surface exposed to the sun, turning over very frequently; or it may be accomplished by baking in a pan over a hot fire.

5. As a final check on the correction, and to adjust the proportion of cement, make the corrected material into a plastic mortar, basing the amount of cement upon the indication of the weight-void table, examine closely. If there is a proper filling of voids, the mixture will appear smooth and dense, with no exposed aggregates or visible voids. If, by the roughness, a deficiency in cement is implied, make second a determination, increasing slightly the proportion of cement; and so forth, until the indication is satisfactory. This proportion may then be adopted for the mix.

6. A determination as above should always be made, regardless of the theoretical indication of proper proportions, which is after all only a guide post, and the appearance of the mixture made the basis of the working proportions. A little practice will soon enable one to hit upon the best proportions speedily in any case. There is nothing equal to the trained judgment—conceived in sound principle and bred in honest practice—in the manufacture of concrete.

Determination of Chemical Fitness:—There should be a determination to ascertain whether the aggregate contains injurious ingredients, such as would react with the cement and induce failure. This may be made by:

1. Carrying on a parallel set of tests with the cement testing, making the same number of test pieces with the sand in question as with the Standard Sand. A very convenient basis

of comparison is thus afforded. The comparison may be made more interesting by making separate series with the sand natural and corrected. Then a further substantiation, or check, on the efficacy of the correction will be afforded.

2. The effect of the sand may also be determined, less accurately but more quickly, by mixing with a sound cement of known properties, using for this purpose proportions about 1-1, observing the setting qualities and the behavior in boiling water. Undue retardal of the set, or manifest indication of unsoundness, will signify that the sand probably needs washing. Let a sample of it be carefully washed and a retest made. This will prove or disprove the presence of injurious matter, and determine whether washing shall be required. This determination should always be made even if the first determination is made.

Direct Determination of Voids:—The gauging of voids by a weight table may not always be satisfactory or it may be desired to ascertain the weight from the voids, in which cases a direct determination may be made conveniently as follows:

Method No. 1.—Required, two vessels of equal capacity, and a good spring balance. The common buckets found on work will be satisfactory.

(1) Weigh the buckets empty and record the average weight. They should of course weigh very nearly the same.

(2) Weigh a bucket brimming full water.

(3) Weigh the second bucket filled to the brim and leveled off with dried and compacted aggregate (sand, gravel, or broken stone).

(4) Pour the water from bucket No. 1 into bucket No. 2 until the material will soak up no more. Add gently so as not to enclose air, and be careful not to spill any of the water. A small piece of hose used as a siphon will solve the problem of transferring the water without loss. Weigh both buckets.

(5) Then, percentage voids equals,

$$100 \times \left(\frac{\text{Weight bucket No. I partly empty} - \text{weight of bucket empty}}{\text{Weight bucket No. I full water} - \text{weight of bucket empty}} \right)$$

$$= 100 \times \left(\frac{\text{Weight water in voids}}{\text{Weight equal volume water}} \right) \quad (\text{I})$$

Check:

$$100 \times \left(\frac{\text{Weight bucket} + \text{aggregate} + \text{water} - \text{weight bucket} + \text{aggregate}}{\text{Weight bucket full water} - \text{weight empty bucket}} \right)$$

$$= \text{Per cent voids} \quad \dots \dots \dots \dots \dots \dots \dots \dots \dots \dots \dots \dots \dots (\text{I})$$

(6) The specific gravity will equal,

$$\frac{\text{Weight bucket aggregate} - \text{weight empty bucket}}{\text{Weight bucket water} - \text{weight empty bucket}} =$$

$$\frac{\text{Weight aggregate}}{\text{Wt. of equal volume water}} = \text{S. G.} \quad \dots \dots \dots \dots \dots \dots \dots \dots \dots (\text{II})$$

(7) The solid or voidless weight equals,

S. G. × 62.5 (weight 1 cu. ft. water).............................(III)

Method 2.—Required an open watertight barrel, fitted with a stop-cock at about the mid-point, a measuring box containing just 1 cu. ft., a good spring balance, and a bucket.

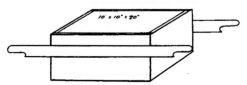

Fig. 61—Sketch Showing Measuring Box for Field Determination of Voids. Note That This Contains 2000 Cubic Inches, Instead of the Required 1728 Cubic Inches.

1. Fill barrel to level of stop-cock with clean water.

2. Measure out 1 cu. ft. of the aggregate in the measuring box, compacting by shaking. The material should be first relieved of its moisture.

3. Introduce the material into the barrel, adding it very gently and stirring the while so as to release all air.

4. Open the stop-cock and catch the overflow; weigh same.

5. Voids equal,

$$100 \times \left(\frac{\text{Weight 1 cu. ft. water} - \text{weight volume displaced}}{\text{Weight 1 cu. ft. water}} \right) =$$

$$100 \times \left(\frac{62.5 - \text{weight displaced}}{62.5} \right) = \text{Per cent voids}............(1)$$

6. Specific gravity equals.

$$\frac{\text{Weight 1 cu. ft. water}}{\text{Weight volume displaced}} = \frac{62.5}{\text{Weight volume displaced}} = \text{S. G.}.....(II)$$

7. Weight solid, or voidless equals,

S. G. × 62.5 = Weight 1 cu. ft. solid material ....(III)

8. This method may be utilized to compare several grades of material, that grade evidently containing the least voids which displaces the largest volume of water.

9. It may also be utilized to determine proportions, measuring the displacement of several differently combined materials, and selecting that which gives the maximum displacement. In all these determinations no more than enough material should be used than will just stop short of the outlet, and a piece of very fine screen tacked tightly over the same so as to keep in the finer material.

Determination of Moisture:—The amount of moisture present in an aggregate may be determined as follows:

1. Find the weight of 1 cu. ft. of compacted material in its natural condition.

2. After thorough drying out, weigh again.

3. Difference between weight moist and weight dry equals weight of moisture present.

4. Per cent moisture equals,

$$100 \times \left(\frac{\text{Difference in weights}}{\text{First weight}} \right) = \text{Per cent moisture by weight} \ldots \ldots (1)$$

5. To get per cent moisture by volume, which represents the proportion of voids in the material occupied by water,

$$\frac{(\text{Weight cu. ft. material})}{62.5} \times (\text{Per cent bv weight}) =$$

$$\frac{(\text{Weight No. 1})}{62.5} \times (1) = \left(\text{Per cent moisture by volume} \right) \ldots \ldots (2)$$

where "Wt. of Material" is that found "moist and compacted", that is Wt. No. 1.

6. Having the per cent moisture the void determinations previously made, may, if the materials have not been dried out before the determination has been attempted, be corrected for "moisture-voids". The net result is the per cent "air-voids", or the "absolute-voids".

E. g., "total-voids" found to be 40 per cent; "moisture-voids", by volume, 5 per cent; then "absolute-voids" equal 40 — 5 = 35 per cent.

Determination of Matter in Solution and Suspension:— Matter is said to be in solution when it is not apparent in the water, as for instance sugar or salt; it is said to be in suspension when it floats around in a finely divided condition, darkening the water, as e. g. clay particles. It may on occasion be necessary to determine the per cent of one or both of these. For this purpose soft, pure water is required, preferably rain or distilled water, which shall contain no matter already in solution.

1. Apparatus Required: Several buckets or pails; a spring balance; a fine wire cloth strainer; some thin blotting paper, such as is known to the chemist as "filter paper"; a bit of ½ or ⅜-in. hose for a siphon; and three wash-boilers, or equivalent vessels.

2. Find weight of 1 cu. ft. of material dry and compacted, or when corrected for moisture voids.

3. Wash material with the clean water at ordinary temperature until effluent is no longer discolored, saving all the wash water. Operation may be best accomplished by stirring the material to be washed into the wash boiler partly filled with water, siphoning off the water into another boiler alongside. The wire-cloth strainer should be interposed at the end of the siphon in order to intercept the finer particles of grit that may have been sucked up. These should be saved and returned to the first boiler after the washing has been completed. The dirt in suspension will pass through the strainer. What matter is in sus-

color of the water standing on top of the sediment that practically all matter in suspension has been dropped, the clear water should be removed by siphoning into a third vessel, interposing this time the strainer covered with a layer of the filter paper. This will intercept what may have been sucked up or is still in suspension. It should be saved and returned to the first residue. Then,

4. Dry the first residue by baking over a fire at low heat, weigh very carefully and record. It should be remarked that each vessel used, in which material may chance to be weighed, should previously be weighed empty and the weight recorded and marked upon it plainly. Then the net weight may readily be obtained. The residue may for convenience be transferred to a smaller vessel for drying.

5. Evaporate by boiling the final effluent (after suspended matter has been removed) to residue. What matter—if any—present in solution will then be thrown down and remain as "residue" after all the water has been evaporated; this is Residue No. 2. Again dry and weigh very carefully. It will be strongly advisable to weigh the residues on delicate scales if at all accurate results are to be obtained, as the amount of residue, especially from evaporation, may be so small as not to be sensible to an ordinary balance. If a drug store is handy, the proprietor may, no doubt, be importuned to do this weighing for you.

$$100 \times \left(\frac{\text{Weight final residue}}{\text{Original weight}} \right) =$$

$$100 \times \left(\frac{\text{Weight No. 5}}{\text{Weight No. 2}} \right) = \frac{\text{Per cent soluble matter}}{} \quad \dots\dots\dots\dots (1)$$

6. Then, per cent matter in solution equals,
If the material under examination has been found to affect the setting qualities, or the soundness, or the tensile strength injuriously, and the soluble matter is in excess of 1 per cent, same should be saved in entire and sent to a competent chemist for analysis.

7. Also, per cent matter in suspension equals,

$$100 \times \left(\frac{\text{Weight first residue}}{\text{Original weight}} \right) =$$

$$100 \times \left(\frac{\text{Weight No. 4}}{\text{Weight No. 2}} \right) = \text{Per cent matter in suspension} \dots\dots\dots (2)$$

$$= \text{Per cent clay or loam.}$$

8. In case it is not essential to know the matter in solution, the wash water after the suspended matter has been removed, may be thrown away, eliminating operations No. 5 and No. 6.

9. The per cent clay or loam may also be obtained by sifting, what material passes the No. 80 sieve being classified as clay or loam, but the method just outlined is preferable, as in the several siftings necessary to separating out the finest material,

pension will finally settle to the bottom of the second boiler, this is the First Residue. When it is apparent by the clean much of it may be lost.

10. These determinations may also be made with much smaller quantities of material, but then the weighings must be more refined and the results in any event will not have been so reliable. The drying of residue is a delicate procedure and requires great care. If the operation is allowed to go too far, the material may be sublimated, that is driven off as extremely fine dust, and the determinations thus vitiated. It were better to have this done by an experienced chemist, making all of the operations except the drying and weighing of the residues, which should be carefully saved, by placing in a jar or stoppered bottle, and sent to the chemist.

Practical Notes on Testing:—1. In making tests, work with as large samples of the material, collected from as many different parts of the pile, pit, bank, or beach as possible, and the several small samples well blended to form the large or working sample. Then one may be assured of a fair average determination. So also in the determinations of weight, voids, et cetera, at least 3 separate and distinct trials should be made and the average struck; and, if any single determination is in error, it will then be apparent and may be discarded and a new one made. In this manner variations in the material and possible errors in manipulation may be averaged and fairly reliable working data obtained. A separate note-book should always be kept for the express purpose of recording all the data of tests. Dates of the month, time of day, weather and temperature conditions, et cetera, should also be herein recorded, as well as the name or names of the persons concerned in the determinations. It will be well also to preserve in glass jars or bottles samples of all the various materials used or considered for use, same being properly labeled. If a separation into grades has been made, corresponding amounts of each grade should be likewise preserved for reference. A very convenient way of doing this is to place the grades in sequence of size into a jar or wide-mouthed bottle, interposing between each layer a thin slice of cork or a disk of tin cut to the diameter of the vessel. Then at a glance the composition of the material may be comprehended. If residues have also been obtained, samples of same may also be added to the collection. Alongside a similar vessel containing the material in the natural condition will afford an interesting and instructive comparison. If, after the purpose of the determinations has been served, it is desired to preserve the visible results for permanent reference, photos of the various vessels may be taken and filed.*

Practical Notes on Selection of Sand:—1. A complete mechanical analysis of sand, involving a separation into grades,

---

*Compare description of a "Cement Record" in Chapter VI.

correction for deficiencies, and recombination as used, in order to produce a blend of maximum density requiring the minimum of cement, should be made for:

(a) All large work of whatever nature, when the saving possible in amount of cement alone will be found to completely justify the additional expense and trouble.

(b) All particular work, such as water-tight concrete and concrete exposed to sea-water, or strong chemicals, or oils.

(c) But not necessarily for small jobs, or jobs of minor importance, when as a rule it will be found more economical to use an excess of cement and let it go at that.

2. If inconvenient to secure a sand sufficiently coarse, or to import suitable material for correction of blend, a fine sand may be used without especial danger, *provided* the *mixture* is made *excessively rich* in *cement,* and it is *not* to be *used* in *sea-water,* where *tests* have *demonstrated* that *fine sand* is *distinctly detrimental.* Provided also the fine sand is not found to affect injuriously the setting qualities of the cement.

3. For water-tight concrete, an excess of fine material, even silt, is desirable.

4. Coarsely graded sands are best for rich mortars; finely graded for lean mortars.

5. Sands with large excess of very coarse or very fine material require large excess of cement.

6. Stone dust and finely divided clay or silt may be actually beneficial in amounts up to 10 or 15 per cent, and in cases in excess of this; loam or vegetable dirt, contrarywise, even in minute percentages, may be detrimental and its presence in general require the washing of the sand. Fine matter should be uniformly distributed in order not to be harmful.

8. Broken stone screenings, if of quartz, trap, basalt, granite, or hard limestone, give, in general, as good and sometimes better results than most natural sands. A blending of a natural with an artificial sand will sometimes prove better than either one alone. Thus, it will not be necessary to remove the dust from crusher run, unless indeed it is found congested in pockets, when it should be removed and used separately.

9. That sand, in general, is best which is heaviest; or made into a mortar of the stated proportions has the least volume

10. The better graded a sand or a gravel the more dense, and is at the same time the smoothest appearing; or in the tensile tests gives the best results.
strong, and impermeable, and the less the amount of cement required to produce a stated result. The most efficient sand is thus the most economical.

11. More coarsely graded sands are desired for top-finishing or surfacing than for concrete proper; more finely graded

sands for small coarse-aggregate than for large; but uniformly graded sands are in all events preferable and desirable for all classes of work, and that grading up to the coarsest size makes densest, strongest, and most impervious mortar or concrete.

12. The proportion of cement in any case may be reduced by substituting for it a proportion of finely ground material, such as powdered clay, stone dust, or any impalpable insoluble material. The density will not thereby be decreased nor will the strength, in the case of lean mixtures, adversely affected; but in the case of rich mixtures, in excess of 10 per cent substitution will begin to depreciate it. Removal of excessively fine material in aggregate and substitution therefor of an equivalent in cement will increase the strength somewhat. The less the density the more fine stuff may be substituted for the cement without harm.

13. To fill all voids in aggregate, there will be required approximately 1 lb. of cement and 3/10 lb. water for each per cent voids per cubic foot of material. (Combined and compacted.)

PROPERTIES OF COARSE AGGREGATES.

Definition:—Coarse aggregates, by conventional division, embrace all particles sized above ¼-in., the maximum size being fixed by the practical conditions of usage.

Maximum Size of Coarse Aggregate:—In general, as large sizes as possible should be employed, since the larger the individual pieces of coarse aggregate, if they be of good stone, the denser, stronger, more impermeable the concrete, the less cement required per cubic unit of concrete and therefore the more economical the mix. A limit is, however, fixed by the following considerations:

(a) *Effect of Consistency of Mix:*—If the mix is fairly dry and is to be placed in thin layers and well tamped, stones as large as 3 in. are perfectly workable, provided there is no reinforcement to interfere. If, however, a wet-mix,—as is the general and approved practice at the present time,—the limiting size will be about 1¼ to 1½ in., which is about as large as will be found to incorporate freely with the rest of the mixture and not separate out during the sequence of handling prior to final deposition. If larger stone than this be used they should be considered as "bulk-swellers" simply, and not as integral parts of the graded aggregates. They may then—if not too large—be intermingled with the rest of the ingredients during the operation of mixing, or they may be added subsequently as the material is deposited. In this case the concrete proper may be considered as a coarse mortar, and the final mass a "machine", "plain", or "cyclopian-rubble", depending respectively whether the "bulk-swellers" are added in the mixer, placed by hand, or by derrick. We have seen that the larger the individual pieces of coarse aggregate and the more of them, up to the point where the mortar

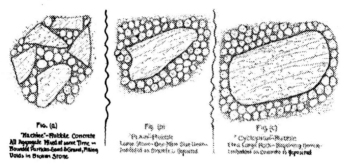

FIG. (a)
"Machine"-Rubble Concrete
All Aggregate Mixed at same Time —
Rounded Particles-Sand &Gravel, Filling
Voids in Broken Stone.

Fig. (b)
Placed-Rubble.
Large Stone-One-Man Size Hand-
placed as Concrete is deposited.

Fig. (c)
Cyclopean-Rubble.
Extra Large Rock-Requiring Derrick-
embedded as Concrete is deposited

FIG. 62—GRAPHIC PRESENTATION OF VARIOUS CONCRETE
MIXTURES.

ceases to be sufficient to neatly fill all voids, the more efficient
and more economical the concrete; however, it is prerequisite that
they be evenly distributed throughout the mass, and no larger
sizes should be used in any case whatsoever than one can be sure
of having properly incorporated. It is also essential that the
pieces be well-rounded, since square, sharp angled pieces, especially rubble and cyclopian size, are likely to occasion cracking, as
in setting the mass shrinks.

(b) *Effect of the Reinforcement:*—Obviously, no stone should
be used sized larger than will pass freely between and around
the various parts or pieces of the steel component of any member.

Bars are packed most closely in beams and girders, and are
frequently also placed in two or more tiers, so that not more than
an inch may separate the individual pieces horizontally or vertically. This condition would then prescribe a limit to the size of
aggregate around $\frac{3}{4}$ in., and in cases to $\frac{1}{2}$ in.

In columns, floor slabs, and walls, it is customary to space
the steel units much further apart and in single tiers only, in
which case there will seldom be any interference, and the maximum size may be fixed according to the first condition. In the
ordinary reinforced-concrete installation, however, it will usually
be found advisable to adhere to the same mix throughout in order not to complex matters.

(c) *Thickness of Member:*—Obviously, no larger sized aggregate should be used than will merge itself freely within the
compass of the member, finer material being required for a thin
wall or slab than for a thick one. In case of intricately molded
designs, to or in any situation requiring the concrete to find
small openings, nooks, and crannies, the material must be super-
fine.

Minimum Size of Coarse Aggregates:—The minimum size, by definition, is ¼ in.; however, if the sand is sized up to ¼ in., the minimum size coarse-aggregate should not be less than ¾ to 1 in. If the coarse aggregates grade down close to the size of demarcation, then there will be a crowding apart of the larger pieces, an excess of cement will be required, and a less efficient and less economical concrete result. It is always better to grade up the fine-aggregates to as coarse as degree as possible, and then use only one or two grades of coarse-aggregate, sized at least four times the maximum size of fine-aggregate, and preferably still larger. Evidently, in general, and in reinforced work especially, only *one size* of *coarse-aggregate proper* will be required. In this case, then, correction of the grading will resolve itself down to correction of the fine-aggregates solely, and the efficiency of the mixture determined on the basis of sufficiency of mortar or fine concrete to fill flush the voids in the large material. Practically, this condition will be realized when on lightly tamping the deposited material with the foot the fine-stuff flushes up, indicating that the voids are just nicely filled.

Laws of Grading:—The laws of grading are in this case of little importance, unless the aggregate is classified all together, when the same rules explained for the fine-aggregates will apply. This will frequently be the situation if the aggregate be a sandy, well graded gravel, or a finely crushed "run-of-crusher". The mechanical analysis may then be carried out precisely as outlined for fine-aggregates, the material being separated into four, five, or six grades, correction supplied for the deficiencies in any grade or grades, and recombination made as used.

Suitable Materials for Coarse-Aggregates:—The question of suitable materials for coarse-aggregates is evidently very important, insomuch as the coarse-aggregates form the burr, as it were,—and the finest, the heart,—the kernel, of the mass, enveloped in a matrix of mortar.

(a) *Coarse-aggregates* should only be of the hardest, most durable, tough, fire and fume resisting material. *Trap and basalt,* being the product of the great smelting pots of nature, are thus the most suitable aggregate, although their weight is often uged against them as an objection. Trap will average 180 to 185 lb. per cu. ft. solid, as against 160 to 170 for granite, quartz, conglomerate and hard limestone. Hence, if trap is sold by the ton instead of by the cubic yard, and the same price is charged as for hard limestone, the economy possible by buying the lighter material will constitute a strong temptation to give it preference over the heavier, if better, material.

(b) *Granite* makes a fairly good aggregate, but has the disadvantage of spalling too readily under the effect of fire, being otherwise excellent in every respect. It should not be used if trap or hard-limestone is available.

(c) *Conglomerate*, or "pudding-stone", which is a sort of natural concrete, also makes fair aggregate, although a good quality is not always obtainable. Very often the particles are not well cemented together. If not, the material is not satisfactory, except it be crushed and used as sand.

(d) *Hard sandstone* is at times a satisfactory, but less desirable, material. Soft or shaly sandstones should not be used. The great objection to sandstones is their lack of toughness and their porosity.

(e) *Hard limestones* and *magnesium limestones* (dolomite) make excellent aggregate, indeed have frequently on test shown themselves to make the strongest concrete. This fact is due, no doubt, to the natural affinity existing between it and the cement, which makes for a stronger union between the materials. Limestone is frequently prohibited on the ground of its lack of durability,—not being a good resistent of either fire or acid-fumes.

Yet, however true this may be of the stone in its natural condition, when encorporated in the concrete mix and for the most part entirely protected from the air by a coating, siliceous sand and cement, it ceases to be attacked readily and becomes instead one of the best materials for concrete. It should not, however, be used for concrete exposed continuously to high-temperatures, like flues or stacks, or in neighborhoods where the atmosphere is heavily charged with acid fumes. Limestone dust is especially serviceable in lending tone to the surface and in adding to the impermeability. Also, being alkaline in its chemical nature, it exerts a beneficial influence on the reinforcement, helping to neutralize the effect of infiltrating acidic moisture or water.

It is to be recalled in this connection that the compounds formed by the cement itself as it hardens tend to revert to the carbonate form, particularly the lime which is the predominant element, so that when limestone is used for aggregate the eventual concrete may become almost solid limestone.

Magnesium limestones have this objection, that the presence of magnesium, with its greater affinity for some of the compounds present or forming than lime, tends to produce a condition of instability, which might result ultimately in disintegration of the concrete; however, the magnesia in this case being thoroughly incorporated in a dense, hard stone, the danger would seem slight and such material, provided no better were available, safe to use; if it were free magnesia or in a condition to dissolve freely from the surface of the stone, then it would be a different matter.,

Marbles are a type of metamorphic limestone, and while they make a very beautiful concrete are not a durable aggregate. They are permissible only for surfacings on which a high polish is desired, and are then in finely crushed form as a part of the surface mixture only.

(f) *Slaty* or *shaly rocks* are not good material for concrete aggregate, and should never be so employed.

(g) *Good rock* will in general show its quality in the manner of breaking; for instance, if it breaks up on crushing or pounding into "bastards", or irregular, chunky particles, with no apparent planes of cleavage, having what is known in geology as a "conchoidal fracture", it is satisfactory; if, on the contrary, it breaks up into more or less regular pieces, flat and slaty-like, it is, as a rule, unsuitable, since such material is more than likely to be soft and brittle, and if not are at any rate unsuited by their shape to incorporate well in concrete. The material known as road-metal, either trap or hard-limestone, makes also good material for concrete. The run of crusher is not objectionable, if the fine stuff is well distributed, since the excessively fine rock known as dust is effective in the mix as increasing the density and the impermeability, effecting also a saving in cement, and, if limy, acting as binder besides.

(h) *Gravels* may be said off-hand to be more generally satisfactory material than broken-stones, since by reason of their ages of exposure to corrosive and erosive agencies, while grinding and rubbing on the primeval beach or under the heel of glacial flows, by which they were made from huge chunks torn bodily off the mountain side or ripped up out of the bowels of the earth, their quality, it may be said, has had a fairly good testing out. Moreover, they are usually pretty well graded naturally, requiring but slight correction to make them an ideal blend—since nature, in the operation of grinding, has sought persistently to bring the particles into closer and closer contact, eliminating the voids. Accordingly, concretes of gravel aggregate are usually harder, denser, more impermeable, and in the long run stronger also, although the indication of strength on short time tests may show a superiority for broken-stone. This inferiority is due, no doubt, to the smooth, well-polished surface of gravel, and their rounded shape, which reduces the internal friction of particles and hence the apparent strength; however, when the cement has attained its full effective adhesiveness the conditions are reversed, and the superior strength of the gravel concrete is then probably due to the more perfect blending of the particles. Another marked advantage of gravel aggregates is the greater ease with which they mix and handle, their rounded corners reducing the friction of contact and allowing the various particles to slip and slide on one another, both in the dry condition and especially in the fluid-mixture when lubricated with cement cream, with the greatest freedom. For the same reason also the danger of voids in the mass, due to bridging of aggregates, is largely reduced. All these considerations make for a marked economy in the labor-cost of every detail of the operation from the feeding of materials to the mixer to the final deposition in the moulds.

The chief objection urged against gravel, excepting that on the score of lower strength on short time, is that the pieces are liable to "pop" under the effect of the flame, especially when suddenly chilled after heating with water. Gravels, however, are no more objectionable on this account than many, indeed most stones.

Gravel concretes are usually lighter in color and slightly heavier, because denser, than stone concretes, except, of course, trap rock concretes.

The material as it occurs naturally will be found to contain considerable fine aggregate, requiring but little if any 'sand to perfect the blend. In cases it may be convenient to remove this excess of fine material by screening, recombining it as used. If from the bed of foul streams, or from a loamy bank, it will need washing. What has been said of cleanliness in regard to sand applies of course with equal force here.

Gravel sometimes contains proportions of soft stone-like shale, and if there is a noticeable amount of such stuff present the material is unsuited for use until separated out. Some failures of concrete have been attributed to the free use of gravels containing such material.

(i) *Gravel-Broken Stone Combination*:—Perhaps the best aggregate for concrete results from the combination of sand, gravel and broken stone. In this case the concrete proper is

SAND — GRAVEL-BROKEN-STONE-CONCRETE
Showing Close Interlocking of Irregular Stone
with Rounded Gravel, increasing Density & Strength

FIG. 63—DETAIL OF GRAVEL-STONE CONCRETE.

really based upon the gravel and the broken stone is "bulk-sweller" only. Tests show that this blend gives both a stronger and a denser concrete than either gravel or broken stone alone. The fact of greater strength, no doubt, is due to the irregular and large pieces of broken stone acting as reinforcement, interlocking the mass in every direction, and of greater density to the sealing of all connecting passages in the sand and gravel by the uniformly distributed pieces of broken stone.

(j) *Cinders,* considered from the viewpoint of fire resistance, are a very satisfactory aggregate, since theoretically they are from the nature of things about as near perfect incombustibles as it is possible to imagine. However, they are open to several objections that unfit them for wide usage. The chief objections to cinders group themselves about their porosity; this makes them light, true, but not nearly so strong as stone, and far less dense. Consequently, cinder-concretes must be figured at a considerably lower crushing strength, and for work requiring a dense, impermeable concrete, are absolutely unsuited. It is to this porosity too, no doubt, that the speedy corrosion of imbedded metal is sometimes found to occur in cinder-concretes after a few years' usage is due. The presence of sulphur also has been ascribed as a cause for this corrosion, but it is more likely due to porosity. A dense concrete of cinder-aggregate may, to be sure, be made by using an excess of cement, but this is expensive and open to other objections besides. Unless this is done, however, cinder-concrete, under the ordinary specifications, which require the screening out of all ashes, will not be a satisfactory material for any reinforced work. It were better that some of the ash be left in, which would help to fill the voids and seal the pores. Professor Woolsen, of Columbia University, who has experimented a good deal with this material, claims that in this way a satisfactory concrete can be made in which steel will not corrode. One great objection to cinders is the great difficulty of getting a good quality. If hard coal, steam-furnace cinders, thoroughly burned out, were always obtainable it would be a different matter, but the majority of stuff offered as cinders is little better than coal-ash, and usually full of unignited particles. These of course soon burn out in event of fire, leaving the concrete correspondingly pitted.

The two chief reasons why cinders as aggregate are interesting are: (1) Their cost, and (2) their lightness, which makes it possible with them to secure a very light, cheap concrete, and, if the cinders are of good quality, of undoubted fire-proofness. Thus, for very many purposes, particularly tall office buildings, apartments and hotels, with skeleton of structural steel, they make an ideal fireproof floor. In this case, too, their porosity is open to less objection, since the floors are seldom if ever wet, the ceilings are heavily coated with plaster and more often doubleceiled, and the top-surface either coated with a dense cement wearing surface or a tight-board decking, so that practically no moisture or acid-fumes can get through to contact with the steel. The chief danger, then, will be from electrolysis. However for fireproofing structural steel work in general cinder concrete seems by far the best material. Frank Gilbreth recently had occasion to wreck a steel frame building fireproofed with cinder concrete, in San Francisco, which had been up for 13 years, and he bore witness that were no signs of corrosion, even where the cinders were in contact with the naked steel. Evidently, then, cinders as aggregate have here a legitimate application. They are also a natural material to use for filler under floors, and for drainage layer under sidewalks and cellar bottoms.

The *Standard Specifications for Cinders* states that "they shall be of hard, clean, vitreous clinker, free from sulphides, unburnt coal, or ashes." It will also be well to remember in using cinders where steel reinforcing is entailed that they must be used with an excess of *fine. mortar* and *very wet*. Also, that they be only for light floor or wall construction and fireproofing purposes, never for footings, foundations, reinforced beams, or heavy floor loadings, or for reinforced concrete structures of the pure type.

Pumice-gravel is a substitute, or rather alternative, proposed by some who are disposed to believe cinders contain too many dangers to be a safe material. Pumice is the product of volcanic action and is a light, spongy, but tough rock. As a light, strong aggregate for concrete it would seem ideal, as its specific gravity is only a half to two-thirds, so that concrete made of it would weigh not more than 75 pounds as against 100 pounds for cinder and 150 or more for good rock concrete. Its great disadvantage is its porosity, requiring the use of an excessive amount of cement to seal the pores. An advantage is its apparent great affinity for cement or lime, which makes for the greater strength of the concrete. Another advantage is that concrete of which pumice is the aggregate forms an excellent nailing material. Tests made by William Challoner (an English engineer), of pumice concrete made of 12 parts of pumice-gravel to one part cement to one part trass (puzzolana matter forming of itself a natural cement), showed a crushing strength, after six months' hardening, of over 500 lbs. per sq. in. and a 6-inch thick slab of it, 3-foot span, carried all the weight that could be piled upon it. Tests made at the Manchester (England) School of Technology, in 1905, on a similar mixture, at 12 months' age, showed an average crushing strength of 537 lbs. per sq. in. Of course, being the product of volcanic action, pumice would be exceedingly high in fire-resisting qualities. Commercially it seems to be about the only available material in place of cinders.

(k) *Miscellaneous Aggregates:* Concrete is veritably a huge mixing pot, in which the odds and ends of otherwise waste material and refuse, so long as it is hard and tough, may be indiscriminately intermingled and by the magic of cement converted into a smooth, dense, hard and durable stone of any form and any dimension whatsoever. Thus, gravel, stone, pumice, cinders, slag, coke-breeze, crunched sea shells, brick-bats, broken glass and crockery, burnt clay, and what not—much of which is otherwise only worthless debris—may be utilized on occasion as filler for concrete, and so made to render efficient service. And, while most of these materials have rather sharply defined limitations, they one and all have a legitimate application. Only first class material, however, is permissible in reinforced concrete or in plain work where great strength, durability, resistance to abrasion and impermeability are necessary. Pumice, cinders, coke, breeze, slag and brick-bats are all open to objection on account of their porosity, requiring an excess of fine material, and especially cement, if good results are desired. Vitrified brick, however, makes a very excellent aggregate.

(1) *Slag*, as a by-product of the iron industry, is much used in some localities where a supply is convenient, and a good quality of gravel or broken stone not so convenient. It is objectionable chiefly on the same account, namely, its porosity, requiring an excess of cement and a very wet mixture. Moreover, there is further objection—that it carries usually a considerable percentage of sulphurous matter in the anhydrous condition (lacking water), and this content, when brought in contact with water or moisture, forms sulphuric acid, which in turn reacts upon the cement, decomposing it and liberating the obnoxious gas hydrogen sulphide (recognizable by the familiar odor of bad eggs). Evidently, should this action take place to any extent within the hardened concrete, the result could scarcely help being disastrous. The writer has seen slag-concrete foundations heave and come right up out of the ground, and upon examination to be found in a badly disintegrated condition.* Great care needs, therefore, to be taken with this material if it must be used, to see that the sulphurous content is not excessive. Slag otherwise is a fair material, making a hard, strong, fireproof concrete of 5 to 15 lb. lighter weight per cubic foot than rock or gravel. It is on this account a cheap aggregate and unless strictly prohibited will be used by contractors because cheap. *It is highly important, if slag or any other porous material is used for reinforced concrete, that the reinforcement be surrounded with fine, rich material.*

In this connection the conclusions of Messrs. Fuller and Thompson in the same paper quoted from the chapter on *Fine Aggregates,* are of interest:

(a) "Concrete of cement, sand and gravel is less permeable than concrete of cement, screenings and broken stone; that is,

---

*This instance has been previously mentioned in this chapter.

for water-tightness, less cement is required with rounded aggregates like gravel than with sharp aggregates like broken stone.

(b) "Concrete of mixed broken stone, sand and cement is less permeable than similar concrete of broken stone, screenings and cement; that is, for watertightness, less cement is required with rounded sand and gravel than with broken stone and screenings.

(c) "Round material, like gravel, under similar conditions, gives a denser concrete than broken stone.

(d) "Sand produces a denser concrete than screenings of similar sized grains.

(e) "A concrete with an angular, coarse aggregate, such as broken stone, is stronger than one with rounded coarse aggregate, like gravel, and the same sand and cement,—although the aggregate produces the greater density—thus indicating a stronger adhesion to broken stone than to gravel. However, if the sand is also angular, like screenings, but with its grains the same sizes as the sand, the concrete with rounded, coarse and fine aggregate is the stronger, probably because of its greater density.

(f) "Aggregates in which particles have been specially graded in sizes so as to give, when water and cement are added, and artificial mixture of greatest density, produces concrete of higher strength than mixtures of cement and natural materials in similar proportions. The average improvement in strength by artificial grading under the conditions of the tests was about 14 per cent. Comparing the test of strength of concrete having different percentages of cement, it was found that for a similar strength the best artificially graded mixture would require about 12 per cent less cement than the like mixture of natural materials."

Weight:—The weight of coarse aggregates is very little different than that of fine aggregates, and what in general was said about the weight of the latter applies equally here. Also, the same weight-void diagram given there will answer here.

The weight of crushed or broken stone will vary as the sand with the relative size and grading, the less the number of gradings and the smaller the individual pieces the lighter the weight per cubic foot. A lot, mostly one size, will average from 75 to 90 lb. per cu. ft., depending on the quality of the rock, trap being, of course, heavier than most other stone; whereas, if well graded, the weight may exceed 100 to 125 lb.

The order of desirability, generally speaking, is the order of heaviness, that is, the heaviest stones make the strongest concrete, but as the heavier stone will go no further and usually entails extra cost for its extra weight, the order of preference, especially on the part of contractors, will be in precise reverse. Therefore, if the heavier stone is actually desired, it should in all cases be rigidly specified to avoid misunderstanding, and to place all bidders on the same basis. Some light-weight aggregates, however, are not really economical, as on account of their porosity the additional cement required with them will very largely offset the apparent initial cheapness.

What has been said of the weight of sands applies with equal force for gravels, since they are but sands on a little larger scale. The average gravel, as it comes from the pit or bank or bottom, will weigh in the neighborhood of 90 to 110 lb., varying according to the degree of grading, the wetness and compactness, even as the sand. The weight-void table on page 197 may be used for gravels. Gravels, as a rule, are heavier than broken stone, since they are better graded and hence more dense.

Solid Weight and S. G. of Various Rocks.

Name.	Weight per cu. ft. solid.	Specific gravity.
Basalt	180	2.9
Brick, Common	100 — 125	1.6 — 2.0
" Pressed	134	2.16
" Fire	150	2.4
Conglomerate	160 — 170	2.6 — 2.7
Flint	162	2.5
Glass, Flint	192	3.08
" Plate	172	2.76
" Common	157	2.52
Limestone	168	2.7
Marble	168	2.7
Porphyry	170	2.73
Granite, Gray	160 — 165	2.6 — 2.65
" Red	165 — 170	2.65 — 2.70
Sandstone	150	2.4
Trap	170 — 190	2.7 — 2.9
Quartz	165	2.65
Sand	165	2.65
Gravel	165	2.65
Gneiss	160 — 170	2.6 — 2.7

Determination of Quality:—Examination with the eye will usually be found sufficient, especially if the examiner possesses some experience with stone, and stone is such a common material that the lad about the street and field knows to some extent what is good stone and what is not.

In addition the following simple tests will be found useful on occasion:

(a) *Fire Resistance:*—Place a fair sample of the material on a shovel or in a pan and hold over a hot fire until the material heats a dull red. Then chill stone suddenly with cold water. It should stand both heating and chilling without softening, spalling, crumbling or checking.

(b) *Acid Resistance:* Treat a sample with strong or concentrated muriatic acid, which may be applied by pouring on the acid or by immersing the sample itself in a vessel of the acid.

Use a glass vessel for this. Violent effervescence, accompanied by rapid crumbling, indicates unfitness. Hard limestones and dolomites (magnesium limestones) are attacked by acid, but less vigorously, than soft limestones and marbles. This distinction will establish a convenient basis for choice between two or more kinds of limestone proposed for use.

(c) *Toughness:* Tap vigorously with a light hammer (a carpenter's will do) four or five times in quick succession. Stone should begin to crush under this impact before it spalls, splits or otherwise fractures. Marked fragility should cause rejection. Some stone, more particularly of a shaly nature, are so tender that they may be broken between the fingers; such, evidently, are not suitable material.

(d) *Porosity:* Take a fair-sized specimen, bake for several hours in a hot oven, temperature above 212 degrees, then weigh carefully, using a delicate balance like a pharmacist's scale. Then immerse in tepid water, leave over night. Remove, wipe off the surplus water with a damp cloth (using a damp cloth and not a dry one so as not to attract moisture from the surface pores) and reweigh. Then, the porosity, or per cent absorption equals

$$100 \times \left(\frac{\text{Weight saturated rock} - \text{weight dry rock}}{\text{Weight dry rock/Its S. G.}} \right) =$$

$$\left(\frac{\text{Weight water absorbed}}{\text{Wt. of equal volume water}} \right) \times 100 = \text{Per cent absorption.}$$

The absorption should not exceed 10 to 15 per cent, and preferably 5 per cent.

The specific gravity, if it be not known from the table of specific gravities, may be found as follows:

Drop specimen of rock into a vessel of known capacity, preferably a graduated glass; measure the increase in volume; then S. G. equals,

$$\frac{\text{Wt. of rock dry}}{\text{Wt. of vol. water displaced}} = \frac{\text{Wt. of rock}}{(\text{Increased vol.}) - (\text{Original vol.}) K} = \text{S. G.}$$

where K is conversion unit, depending on system of weights. If pounds and ounces, and the glass reads directly in cubic inches, then

$K = 62.5/1728 = 0.0362$, which is the weight of 1 cu. in. water in pounds. To be consistent, then, the weight of the rock should be expressed in pounds and decimal, e. g., 2.345; or, if found in ounces, K should be expressed in ounces; K in ounces equals $0.0362 \times 16 = 0.579$, which is the weight in ounces of 1 cu. in. water. It will be especially convenient if the metric system is used, since then 1 cu. centimeter weighs 1 gram, and the weight of the rock in grams, divided by the increase in volume in centimeters, will give the S. G. at once.

It will not ordinarily be necessary to make this determination, except a question arises on account of the evident porosity of the materials, and over possibly the amount of water that should be used in the mix, or an exceedingly dense and impervious concrete is desired.

Having the S. G. the weight per cubic foot solid materials may be obtained by multiplication into the Wt. of a cubic foot of water, that is,

$$\text{Solid weight} = \text{S. G.} \times 62.5$$

(e) The determinations for voids, et cetera, will be the same as outlined for fine-aggregates.

PROPERTIES OF WATER FOR CONCRETE.

General:—In general it may be said that water fit for washing purposes will be suitable for concrete. It should be clean, "soft", fresh, and neutral. The presence of oil, strong chemicals, slime, or other filth or pollution is impermissable. It used to be thought that common salt was injurious to the strength and permanence of the cement but experiments have demonstrated that, in small percentages at least, it is harmless, indeed by some regarded as beneficial, increasing the fireproofness. The ordinary occasion for adding salt is for the purpose of lowering the freezing point, thus permitting work to be carried on at a lower temperature.

Sea Water:—Sea-water is objectionable for gauging mortars and concrete not because of its saltness* but because of the presence of other salts which act upon the cement during the process of setting or hardening. The injurious salt seems to be sulphate of magnesium, and the chemical action that takes place is supposed to be (as outlined in the section on chemical composition of cements) somewhat as follows:

Magnesium sulphate in water reacts with lime in cement forming calcium (lime) sulphate, precipitating magnesia; the calcium sulphate formed immediately recombines with the alumina in the cement to form aluminous-calcium-sulphate, which is a gelatinous compound possessing expansive properties. The net result is, at any rate, the gradual "rotting" of the cement and the eventual failure of the concrete.

If the concrete is protected from the action of the sea-water until the cement has thoroughly set there seems to be little or no action; nor if it be sufficiently dense and impervious (see article on fine-aggregates) it will be materially affected.

The evidence of the action of sea-water on cement is the foamy slime that is produced when cement or concrete is

---

*Common salt to which is due the saltness of sea-water, is sodium chloride, $NaCl$.

placed in it. It will be found to rise to the surface after every deposital of fresh material and also to coat masses of concrete just laid. Failure to thoroughly remove this slime from the surface of each layer before adding another will result in general in a plane of weakness in the structure, for this slime interrupts the adhesion, furnishes a ready passage through the structure to the sea-water, and thus constitutes a serious danger.

The sea-water may not be wholly to blame. The presence of any unsoundness or an abnormal alumina content seems to encourage the attack very materially, and the use of excessively fine sand, increasing the porosity of the mixture, also is detrimental. It is safe to say that with a thoroughly sound and well-aged cement, having low alumina content and preferably none (being then an iron-cement), the effect of the sea-water will not, if the action is as indicated, be serious; and further, if the grading and intermingling of aggregates be properly done, producing a mass of maximum density and impermeability, the effect should in any event be *nil*. The use, too, of finely divided colloidal substance, as for instance, clay or finely ground hydrate lime, contributing to the filling of voids, might be of benefit.

Salt Water:—Salt water up to 10 per cent solution seems to be harmless, tending only to delay the setting but not weakening the ultimate strength. The following table, taken from Falk's Cements, Mortars, and Concretes, illustrates the effect of salt in the water used for gauging cement mixtures:

Age.	1 : 2 briquettes. Tensile strength in lbs. per sq. in., gagede with		1 : 3 briquettes.	
	Fresh water.	*Salt water.	Fresh water.	*Salt water.
7 days	236	126	112	68
1 month	289	231	183	131
3 months	414	294	268	215
6 months	549	424	335	266
9 months	554	452	351	301
12 months	572	576	458	413

*The solution was about 10 per cent.

If salt is added to the water, great care should be taken to see that it is thoroughly dissolved and that the solution is not overdosed. If the effect is not bad on the strength of the concrete it is on the looks of it. Too much salt, or salt in lumps, will eventually work its way to the surface, showing in blisters and unsightly stains. It should never be added in lumps directly to the mixing water but in the form of brine.

Fresh water:—Fresh water may contain mineral matter in solution that will cause the water to have as injurious effect as sea-water. Because water comes from an inland lake or stream is not proof-positive that it is suitable. Its effect on the cement to be used should always be learned.

Temperature.—The temperature of the water used in gauging exerts a marked influence on the rate of set, which seems to vary directly with the rise from the freezing point, being virtually nil at 32 degrees F., normal at about 70 degrees, and almost instantaeous at the boiling point. This fact is made the basis of the accelerated tests for soundness, and is applied commercially in the molding of cement building tile. In the latter process an excessively soft mixture is poured into metal molds which are so arranged that steam may be introduced on all sides at once. In 8 to 10 minutes the set progresses far enough to permit of the removal of the tiles, and three days subsequently in a hot room, in which steam is constantly admitted, results in a hardening equivalent to 30 days at ordinary temperature.

An important point about the temperature is this: that, as the rate of setting increases in direct ratio with the rise in temperature from the freezing to the boiling point, so also does the degree of formation of colloidal rather than crystalline compounds. The strength and impermeability being in direct proportion to the relative formation of colloids, the closer the temperature of the entire concrete mixture approaches 212 degrees F., the better, theoretically, the concrete. However, as evaporation also increases very rapidly with the rise in temperature, the higher the temperature at which the mix is run, the more water it will require and the greater the need of protection from external heat—e. g., the sun's rays—until the process is complete. The reason why concrete mixed and placed in hot weather is often inferior—when it should be superior—is because proper precautions are not taken to minimize evaporation and provide the required excess of moisture to the surface. Heat, though a *desideratum,* should not be used *ad libitum.* It is always necessary to ensure a slow enough rate of setting to permit of the reasonably free handling of the mixture through the necessary steps preliminary to final deposition. The average working temperature, therefore, should always be under 100 degrees F. and preferably never above 70 to 80 degrees F. If hot water is ever to be added to the mixture, it should never be brought in direct contact with the cement, but first mingled with the aggregate, which will sufficiently distribute the heat and lower the average temperature to a safe working degree.

However, in actual construction, while at temperatures between freezing point and 50 degrees F. the set does not take hold as soon as if the temperature were in the neighborhood of 70 degrees, yet the heat generated internally by the incipient chemical action soon warms up the mass to a degree where normal rate of setting ensues. Hence, little attention need be paid to the temperature of the water unless both it and the atmospheric temperature are around freezing point, when it is advisable to heat the water. This may be conveniently done by exhausting from the boiler into the water tank.

A freezing temperature does not injure, simply delays or postpones the setting of the cement, and unless an alternate freeze and thaw occurs before the material takes a hard set—which alteration will probably effectually ruin it—the only disadvantage will be that due to the loss of time. Some discussion of this was given in the paragraphs on *Set*. The following table, compiled from the Watertown Arsenal Report for 1901, gives a valuable indication of the effect of freezing temperatures:

Brand.	Length of time at 0° Fahr. Months.	Days.	Subsequent Length of time at 70° Fahr. Days.	Compressive strength in pounds per sq. in.
Star Portland Cement, 1 : 1 Mortar......	..	5	7	846
	..	14	7	1,000
	..	21	7	1,010
	..	31	7	981
	2	..	7	981
	3	..	7	1,010

Note: Result on one brand of cement only is given, as it is typical of the results reported. Test specimens were 2″ cubes.

It is evident that no setting takes place while the water in the mixture is frozen, since no matter the length of time in that condition the strength obtained in 7 days subsequently is practically the same in all cases.

Some slight setting does, however, undoubtedly take place in actual work, and is due, no doubt, to evaporation, a formation of water vapor taking place on the surface of ice in the focus of the sun even in the coldest weather. This escaping vapor, part of it, is taken up by the cement and compelled thus to initiate faint chemical action. The operation is akin to that of washed-clothes which are said to "freeze dry", and in the case of thin layers like plaster coats a precisely analogous action does take place, and the plaster takes hard-set at temperatures below freezing. It will not, however, be safe to rely upon such action as an aid at all in the ordinary concrete installation.

Quantity of Water Required:—The quantity of water required in gauging or mixing concrete or mortar will depend chiefly upon the density of the combined aggregate. It will also depend upon the porosity of the aggregates, the fineness of the cement, the amount of dust or excessively fine stuff present, and the amount of water already contained. It will also be influenced slightly by the temperature, since the solubility of liquids increases in general with the rise in temperature; this is not always true, as some salts in solution are thrown down by a rise in temperature; but as it applies to concrete mixtures it is true. Consequently less water will be required to impart the proper consistency to the mixture in warm weather than in cool, and in cold weather less heated water than water at natural temperature. This difference cannot be expressed by percentages; it is more a matter of judgment. However, it is a point worth remembering, since on a cold day one may choose to heat the water used in mixing in the early morning but not in the later day, and having adjusted the quantity of water at the outset wonder why, later, the mixture is too stiff.

If, however, the mixability is enhanced by the rise in temperature and thus the relative amount of water required decreased, on the other hand the evaporation increases so rapidly with a rise in temperature that usually an excess of water is required for high temperatures over and above that required for lower temperatures, the greater provision must be made, by supplying surplus moisture externally, to offset the increasing loss by evaporation.

Lean mixtures require less water than rich ones; large sized aggregates less than finer sized; dense aggregates less than porous; mixtures free from dust and fine material, besides the cement, less than dusty, loamy mixtures; and moist materials less than dry.

Gravel mixtures will in general require less than broken stone, since they usually contain already more or less moisture, are less porous, and contain less dust. Fine sand mortars will require more than coarse sand, as the voids are greater, but on the other hand they are usually wetter naturally, being because of their fineness more retentive of moisture. The greater the proportion of cement the more the water, since the main function of the water is to put the cement into solution and to hydrate it.

The minimum water, obviously, will be required by the best graded mixture, containing the minimum of cement necessary to produce a given result.

Too much water giving rather too thin a mixture is preferable to too little water giving too stiff a mixture, especially in reinforced-concrete, since the surplus water—if not so excessive as to cause the mixture to separate badly—will come to the top, where it may be bailed or swept off, or allowed to evaporate.

The amount of water, too, will vary with the fineness of the cement, the finer the more water, but this makes less difference in concrete mixtures than in neat cements and rich mortars.

The chief office of the water, we have seen, is with the cement to form a thin liquid which fills all the voids in the aggregates. The amount of water thus required will be such as will fill to saturation all the voids in the cement. *One sack* of Portland cement may, for the sake of proportioning, be assumed to hold *one cubic foot* and to *weigh 100 lb.* The *specific gravity* of Portland cement is about 3.1 that is a *solid cubic foot* will

241

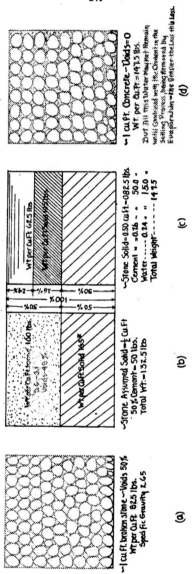

FIG. 64—GRAPHIC PRESENTATION SHOWING PROPORTIONS OF
MATERIALS FOR CONCRETE.

weigh about 195 lb. Hence, in 100 lb. of cement there is 100/195 or 52 per cent of solids and 95/195 or 48 per cent of voids. To put, therefore, 100 lb. cement in solution, there will be required 48 per cent of an equal volume of water, that is 48 per cent of a cubic foot of water, or 48/100 of 62.5 lb. equals 30 lb. water. Or 3/10 of a pound of water will be required for each pound of cement. This will be convenient to remember. Also, for each per cent voids in an aggregate one pound of cement has been found to be required (see "Proportioning Cement to Aggregate"). Hence the rule may be stated *"3/10 lb. water to each per cent voids in aggregate per cubic foot of dry and compact material."*

To this amount an excess of, say, 10 per cent should be added for good measure. The condition of the aggregate as used (in regard to moistures), and the temperature at which the operations are to be conducted, need also, on occasion, be taken into consideration.

These facts of proportioning have been represented on the illustration herewith, which is self-explanatory, and platted on the accompanying diagram for ready reference. The curve of voids and the curve of cement have also been included on this diagram for convenience. It will be seen that the greater the density, that is the more perfect the grading, the less the amount of water required. The great objection to both excessively rich and poorly graded mixture will now be apparent—to wit, that the richer requires more water, and the more poorly graded more also. The condition of the aggregates demands this to satiate the voids, hence the greater the opportunity for loss of weight by evaporation while the cement is gaining its full set, — the greater the liability to porosity, efflorescence, weakening and dusting of the concrete. With a properly graded mixture, of maximum density, the loss of water by evaporation during setting would be a minimum, practically all of the water used would remain in the heart of the concrete to fulfill its purpose, and what of the cement it had in solution would remain in solution and not be drawn with it to the surface and there left as a disagreeable deposit as the water evaporated. The vital necessity—for the same reason—of keeping all exterior surfaces well moistened for as long as possible is also hereby enforced —water must be supplied the surface in order to prevent the water in the interior essential to the full and proper setting of the cement from being attracted to the surface and prematurely evaporated.

Examination of Water for Concrete:—The suitability of water for concrete may be determined as follows:

(1) *Physical Appearance:*—Appearance to eye—it should be clear.

Taste—it should be tasteless.

Smell—it should be odorless.

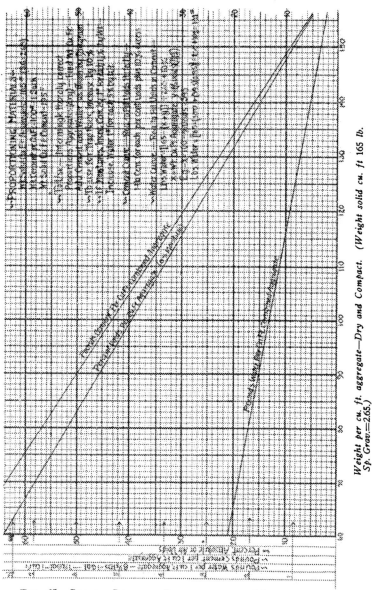

FIG. 65—CURVES SHOWING PROPORTIONS OF MATERIALS FOR
CONCRETE.

Feeling—it should be perfectly limpid, no stickiness, oiliness or grittiness. Try with soap—it should make a smooth, luscious lather.

If water registers O. K. on all these simple trials, it is satisfactory; if suspicious, it may be further examined as follows:

(b) *Acidity or Alkalinity:*—Tests with litmus paper. Litmus is a chemical which indicates acidity or alkalinity by a rapid change of color, responding even as a barometer does to a change in atmospheric pressure. Paper treated with it may be procured at any chemist's. Get both *red and blue litmus paper.* Try with the *blue.* If the paper remains blue on immersing in the water, then the property is either neutral or alkaline; if the color changes quickly to a red, then acidic, which will call for the water to be further examined for its effect upon the setting qualities of the cement. Strong acids eat cement. Faint acids, especially the acid likely to be present in ordinary waters, carbonic-acid, are practically harmless. Indeed, carbonic is a positive benefit, helping to convert the lime present into carbonates. The effect upon the setting qualities should in any event—if the indication is red—be ascertained. Usually, with fresh litmus, if there is a dangerous amount of water present the change of color will be very rapid—if slow and faint, then it is fairly safe to assume that the amount of acid is negligible. Suppose the water to have been tried with the *red* litmus. If the color changes very quickly to a blue, then strong alkali is indicated. If this is the case, the effect upon the setting qualities should be again learned. Alkalinity is in general less dangerous than acidity, indeed alkalinity is the natural property of cement itself. Hence, unless the indication is very marked, and is substantiated by other tests, especially by the effect upon the set, the indication may be disregarded. Some river waters, especially of streams into which the waste of chemical and iron industries empty, will be found to contain injurious percentages of acids, usually sulphuric; and in some cases strong alkali. Such evidently will not be suitable. If the indication is virtually neutral, that is if it is faint for both the blue and the red litmus, then the water may be assumed fit.

(c) Boil a definite quantity by weight (or volume) to residue. Calculate the percentage by weight of solid matter. If in excess of 2 or 3 per cent, the water should be submitted to a competent chemist for analysis, and the nature of the matter in solution learned. The short test for effect on the set is not very positive, since ingredients may be present that will only develop unsoundness after considerable time. Less than 2 or 3 per cent may with fair assurance as to safety be disregarded. And if manifestly silt or fine grit it is in greater percentages negligible.

(d) The best test of water will be to use for gauging a

comparative set of pats and briquets when testing the cement itself. Any marked difference negatively will indicate unfitness. While this indication is by no means absolute, for the reason given in (3), it is nevertheless the most positive test that can be made, and usually it will be safe to accept its witness as final.

(e) As a simplification of (4), and as a convenient method in any case, try the effect of the water proposed for use upon the setting qualities of the cement, and also ascertain by making a pat of a known-to-be sound cement its effect upon the soundness, using the boiling test to expedite matters. Any strong quality of unfitness will be sure to show in one or the other of these two determinations.

APPENDIX A.

AN ANALYSIS OF THE ACTION OF CEMENT*

Dr. W. Michaelis Sr., in a paper read before the German Portland cement manufacturers at Berlin, Germany, March, 1909, which has been set over into English by Dr. W. Michaelis Jr., and published in booklet form by "Cement and Engineering News" of Chicago, presents in some detail his theory with regard to the hardening of Portland cement. This treatise is the culmination and digest of 45 years of study, experiment, and investigation, and the author expresses confidence that he has at last succeeded in unveiling the mystery which has perplexed and confounded scientists for over 150 years. If so, the cement world is deeply indebted to Dr. Michaelis, for his better part of a lifetime's painstaking devotion to this task, and congratulate him at last upon its apparently successful termination. We are of the opinion, after a careful reading of the paper, that, if a positive solution has not been reached, at any rate considerable light has been thrown on the subject, and we feel that we now much better understand what takes place in the phenomenon of the setting and hardening of Portland cement.

The commonly accepted theory with regard to the hardening of Portland cement, upon being admixed with water, is that it is the result of a crystalline growth, by means of which the constituent parts of the cement, broken down by the water and re-associating with it to form new compounds, take on a crystalline structure which enlocks the included particles of inert matter, sand or gravel, in one continuous grip. The exact composition, however, of these crystalline compounds, has continued to baffle investigation, and therefore the theory verification.

With this explanation Dr. Michaelis takes issue, maintaining in lieu that the reaction is almost entirely, and essentially, of a colloidal nature, or briefly that the cement when admixed with water becomes in effect a mineral glue. The reason, he maintains further, why analysis has failed persistently to unravel the mystery is because no definite compounds are formed, but that the reaction is of a very complex and heterogeneous nature and continues indefinitely, the final eventual probably being the conversion of all lime into carbonate, and all silica, alumina, and iron into crystalline hydrates.

Dr. Michaelis' theory of the reaction is, briefly, this: That the water, upon being brought into intimate contact with the

---

*By Mr. Porter. This is reprinted from Concrete Engineering, November, 1909.

grains of cement begins immediately to dissolve and disintegrate them, associating itself with whatever loose or liberated lime there is. The solution becoming thus speedily strongly alkaline, limy, begins to attack the silica alumina, and iron, forming insoluble colloidal hydro-silicates, aluminates, ferrates, and probably also the simple hydrates of these same substances. This reaction affects only the surface of the cement grains, the hearts remaining as yet intact. The result is that almost immediately each little grain becomes enveloped in a gelatinous coating which is the colloidal hydro-silicate of calcium, or as it is also known, **hydro-gel.** This gel, by evaporation without and absorption within,— giving up part of its water to the atmosphere and to the enclosed grain,—coagulates, becoming finally a dense, opaque, water and air-proof solid. The set proper is said to be taken, or the coagulation or hardening process to have begun, when the water vanishes from the surface of the mixture and it assumes a dull face.

There is no apparent end to this process. The heart of the grain continues to absorb water from, and surrender calcium to, the hydrogel, and the hardening thus to go on, for an indefinite period. Hence the phenomenon of the gradual and progressive increase in the strength of Portland cement mixtures throughout a number of years, and also of the retension of hydraulic properties by hardened cement upon regrinding.

The cement grain contains free lime to surrender because, in the calcining process, it is virtually impossible to effect a perfect combination of all the lime, with the result that Portland cement is always heavily over-limed. What Portland cement clinker really is, is a solid solution of lime partly dissolved and partly enclosed in fused silicates, or puzzuolanic matter. Water dissolves this lime, the lime water in turn dissolves the silica, forming the insoluble colloidal hydro-silicate of calcium, or hydrogel.

Hydrogel is at first only semi-impermeable to lime water,— that is it allows water to percolate but retains the lime. Hence the phenomenon of the continuous absorption of lime from the interior of the cement grain by the hydrogel, and the surrender of the water of the latter to the same, by which exchange the hydrogel grows more opaque, solid, and impervious, and the cement grain is hydrated.

There is undoubtedly some crystalline growth, due to the formation of hydrates of calcium, aluminum, and iron, and if gypsum (calcium sulphate) be present, calcium sulpho-aluminate also. All these compounds, in solidifying from a lime solution, take on crystalline structure, either slender, needle-like crystals or large flat, hexagonal ones, depending on the rapidity with which the action takes place. These crystals are imbedded in the hydrogel like so many splinters of wood, or bristles of brush in glue, and of and by themselves contribute little to the strength of the mixture. Moreover, they are soluble, and if it were not for the enveloping insoluble calcium hydro-silicate, hydrogel, they would speedily be dissolved out by percolating moisture. Hence the hardening of Portland cement, and its strength and permanence, cannot be due to the growth of crystals alone. There must also, and principally, be a formation of hydrogel.

In further support of this contention, the formations of crystals can only proceed in air, by evaporation of surplus moisture which, if it were the true and only reaction taking place, would forego the possibility of Portland cement setting under water—which it does.

Setting and hardening under water can only be due to the absorption, by the undissolved cement hearts, of water from the hydrogel, by which it is coagulated. The hydrogel being insoluble continues to harden in spite of the fact that it is submerged, by surrender of its surplus moisture to the enclosed cement grain. Hence the property of hydraulicity of calcareous cements, and therefore the only satisfactory explanation of the hardening is the colloidal one.

The phenomenon of blowing is also hereby explained. Blowing, as is well known, is a phenomenon peculiar to coarsely ground cements, and has constituted one of the strongest arguments for finer grinding. Suppose the cement to be ground very fine, so that fully 75 per cent. passes the No. 200 sieve. Then, the particle enveloped in the hydrogel being exceedingly minute, the expansive tendency developed by the hydration of the particle, by absorption of water from the enveloping hydrogel, will not be sufficient to burst the envelope but only to make it bear the tighter, and thus to increase the strength of the mixture. However, if the grain be large, the strength of the enveloping gel may not be sufficient to resist and blowing ensues, which of course disintegrates the mass. Hence the vital importance of fine grinding.

Dr. Michaelis does not draw the line sharply between the colloidal and the crystalline states, considering them phases of one another. If the solution be only slightly over-saturated, and the formation of solids is consequently slow, large, well-formed crystals are the result. If moderately over-saturated, and the formation of solids is hence more rapid, slender, needle-like crystals result. An example of this is the crystallation of gypsum, which from a slightly over-saturated solution forms large crystals, but from an excessively saturated one, small, needle-like crystals. If the solution be excessively over-saturated, the solidification may become so rapid as to forego the formation of separate or crystalline bodies, and the result is a solid of unbroken structure,—a colloid. The mass, in coagulating, is literally jelly, that is a colloid, having the consistency and physical properties of a glue. In case of Portland cement, the ordinary solution, if not too excessively thinned with water, is in a state of excessive over-saturation; hence the formation of colloids is the natural and inevitable result. And for best results, therefore, an excessive amount of water should not be used, lest the colloidal formation in a measure be retarded and the weaker crystalline structure result.

Explaining the retarding influence of certain chemical compounds, notably gypsum and calcium chloride, upon the setting of Portland cement, these compounds, being themselves strongly crystalline in their tendencies, bring on the formation of crystals from the cement solution, and thus, counteracting in a measure the formation of colloids, act to delay the hardening. So, on the contrary, compounds, like sodium silicate (water-glass), tending themselves strongly to the colloidal condition, counter-act the crystalline tendency and thus accelerate the set. Alkali-carbonates tend also to retard the set, reacting with the lime to form lime carbonate, which prevents the solution from reaching the state of proper super-saturation for the speedy formation of colloids, and retarding the process by dissolving the silica and alumina instead of precipitating them. Retarders should therefore be used with considerable caution, and no more used than absolutely necessary to preserve a safe rate of set to suit the practical demands of use.

By way of support of the colloidal theory, Dr. Michaelis recounts Frederick Ransom's process of manufacturing artificial sandstone, which, briefly is as follows: Pure quartz grains, of various sizes, are soaked in a concentrated solution of sodium silicate, or water-glass. The mass is then molded in the desired form and immersed in a correspondingly strong solution of calcium chloride. The calcium and sodium exchange places, forming, with the help of the water, calcium hydrosilicate, hydrogel, and sodium chloride, or common salt. The former is insoluble and remains intimately intermingled with the quartz grains, cementing them together while the latter is washed out and wasted. The result is a sandstone of great durability and strength, surpassing many times the strength of Portland cement mortar. Natural sandstones are very likely similarly formed.

By way of controversion, Dr. Michaelis experimented with barium cement, made by substituting an analagous line of barium compounds for the lime compounds. The result was a cement of greater strength than the lime cement, because of greater specific gravity, but one that would not harden under water and that would eventually under the influence of water, soften and dissolve. Hence a barium cement would not be a hydraulic cement. This is due mainly to the fact that the hardening process is not colloidal but mainly crystalline, like the formation of gypsum, and hence, like gypsum, soluble. This proves the necessity for a colloidal formation, of the insoluble gel of silicate of lime, in order to secure a hydraulic cement at all.

The explanation of the disappearance of rust from steel imbedded in cement is also made clear by the colloidal theory, the iron oxide entering into combination with the lime and water to form the colloidal hydroferrite of calcium, an analogous compound to the silicate upon which the main reaction depends. This gives a glassy compound enveloping the metal and protecting it from further attack of the negative ions.

Seeking to explain the deleterious action of sea-water upon cement, Dr. Michaelis offered the following: That the hydrogel in coagulating contracts and leaves slight openings here and there, exposing the interior to the ingress of the sea-water. By this ingress any crystals touched would be dissolved. The same would be true of percolating fresh waters, and the result would be finally in any event, a honeycombed mass. But in the case of sea-water, the presence of magnesium sulphate therein constitutes a further agent of disintegration, for the magnesium tends to exchange with the lime in the hydrogel and thus to dissolve the same, penetrating finally to the enclosed cement grain, dissolving it, and forming also with the aluminum expansive sulpho-aluminates, which exerts a further disrupting influence. The only safeguard against disintegration by sea-water is, then, to make the face absolutely impervious, denying water any ingress whatsoever.

Fresh water, penetrating but not percolating, results frequently, after a time, in the closing up of the pores, effecting impermeability where originally there was only semi-impermeability. This result is brought about by the dissolving action of the water upon silica in the presence of lime, whereby a portion of the silica in the cement, and in the aggregates, too, in all probability, enters into the formation of hydrogel. There is thus a growth of insoluble colloidal matter in the pores and interstices which eventually effects perfect impermeability. Particles of colloidal clay, held in suspension in the water, may also contribute to this result. It follows, in all events, that the stability of Portland cements under water is due directly to the formation of insoluble colloids,—in the original formation firstly and secondly in the eventual sealing of the pores by the formation of additional insoluble colloids, which together effectively protect what of soluble or unstable compounds there may be imprisoned within the mass.

There is no practical distinction between the setting and hardening processes, the one being merely the continuation of the other. The process is not complete until all the lime is transformed into carbonates, by the action of the atmosphere or carbonic gas in solution, and all the anions (silicic, ferric and alumininic acid) into hydrates. Until then the chemical changes within the cement do not cease. The solidification of the hydrogel by its continual absorption of lime, is due to its assuming more and more the condition of a carbonate of lime and a silicate of water (hydrate of silica), closely associated. There is no determining just when this process comes to a completion, but the finer the cement the sooner and more positively it would be reached.

If this theory of Dr. Michaelis' is eventually found to be the correct one, then it would seem that the present process of manufacture of Portland cement is in likelihood of radical modification. If lime, silica, and water are the sole agencies necessary to hydraulic action, then these compounds might just as well be brought together directly, forming a liquid cement, and thereby eliminating calcination to clinker and one operation of fine grinding. The operation would then consist simply of bringing together very finely ground silica and a concentrated solution of lime water, resulting in the formation of the flocculant precipitate, calcium hydro-silicate, or hydrogel. This solution would then constitute a mineral glue, and would only need to be used with a proper proportion of lime paste to render it a complete substitute for Portland cement—if the colloidal theory is the correct one. The lime paste is necessary in order to provide a medium for the absorption of surplus water from the hydrogel, forming the solid gel and crystalline hydrate of calcium. This is essentially the basis of the combination of puzzuolana matter, which is fused calcium and silica, with lime paste to form puzzuolana cement.

Or Frederick Ransom's process, using solutions of sodium silicate and calcium chloride, as now employed in the manufacture of artificial sandstone, might also prove available and perhaps more facile.

In any event an interesting and promising line of investigation is hereby opened up.

APPENDIX B.

THE USE OF MINERAL OIL IN CONCRETE.*

The published results of experiments recently made to determine the efficiency of a mineral oil incorporated with the concrete while mixing, seem to promise a wonderful field of development. The account of these experiments follows:

The mixing of oil (mineral) with concrete is very simple. The oil, alkalies and water will form an emulsion, becoming thoroughly incorporated in the concrete. If the concrete is to be mixed by hand, proceed as usual and after the water has been added, the resulting mass turned and raked, add non-volatile mineral oil in proportion of 10 to 15% of oil to the weight of the cement. Turn the concrete with shovels two or three times, raking while turning; the oil will quickly emulsify and become thoroughly mixed in the concrete.

If machine mixing is employed, use a batch mixer, turning a sufficient number of times to thoroughly mix the cement, sand, crushed stone or gravel and water. Then add 10 to 15% or non-volatile mineral oil. Turn again the same number of times as it requires to mix the concrete. The oil will quickly emulsify and become thoroughly incorporated in the concrete.

Oils added to concrete in proportions of from 5 to 15% will slightly delay the initial and final set. Increasing the proportions of oil will further retard both the initial and final set and hardening, but up to 15%, from experiments so far made, it would seem that the retarding of hardening will not be sufficient to cause the work to be uneconomical.

The tensile strength will necessarily be reduced, and with the increasing percentages of oil toughness will be slightly diminished but not in proportion to the increase in the percentage of oil used.

An extremely interesting paper was read at the meeting of the Association of American Portland Cement Manufacturers, at the Hotel Astor, New York, December 15, 1909, by Logan Waller Page, Director of Public Roads, Agricultural Department, Washington, D. C., on the subject of the "Possibilities of Portland cement as a Road Material," in which he described some investigations being carried on by Dr. Allerton S. Cushman in the Laboratory of the Office of Public Roads to ascertain the practicability of mixing semi-asphaltic base oils with Portland cement concrete, with the object of obtaining the desirable properties of both Portland cement and asphaltum. So far only pats and briquettes have been made; the results so far obtained show ample strength for ordinary work; 6 in. cubes will be tested later.

*By Albert Moyer, New York City.

It is believed that compression tests will show greater strength than the usual relation of compression to tension. This is a matter for further investigation, and it is to be hoped that chemists and cement testers will actively take up this work and carry on investigations covering long time periods.

Tensile strain tests should be discarded. Such tests have now been discarded by the German Portland cement manufacturers and compression tests substituted. With the increased scientific knowledge and the consequent better material produced by Portland cement manufacturers, tensile strain tests have become obsolete, and owing to the brittleness and extreme sensitiveness of neat Portland cement, the unscientific methods employed in tensile strain tests, the personal equation involved, tensile strain tests do not indicate the possible load which Portland cement concrete may carry.

Compression tests on cylinders of a size which will cause the area to equal 6-in. cubes should be used. In order that such tests may be standardized and relative, standard sand should be used, and if possible a standardization of gravel or crushed stone. If crushed stone, trap rock should be used, all passing through a $\frac{3}{4}$-in. mesh and all collected on a $\frac{1}{4}$-in. mesh. Mix up cylinders which will theoretically figure maximum density, add varying proportions of oil from 5 to 20 per cent. Also make up another set of cylinders, adding varying proportions of hydrated lime, from 10 to 30 per cent., increasing the percentage of oil with the increase of hydrated lime. The addition of hydrated lime theoretically should permit the addition of a larger percentage of oil, as we thus have a greater emulsifying material. Varying percentages of Portland cement may be used, always keeping the relation between the sand and stone the same, maximum density having been figured. The amount of Portland cement to be increased above that which is required to fill the voids in the sand.

Two months ago the writer made some briquettes and pats with the object in view of ascertaining if the mixture of oil with wet neat cement and mortar would have the tendency of keeping all but the excess water from leaving the wet neat cement or mortar.

Briquettes were made, neat cement mixed with water, the water slightly in excess of that usually required, after which 10 per cent. of oil petrole was added. (Oil petrole is a white, non-volatile petroleum product of about the same consistency of melted vaseline.) Pats were made of 1 part cement, 3 parts sand mixed with water, a little in excess of what would ordinarily be used, after which 10 per cent. of the same oil was added. These pats were about $2\frac{1}{2}$ in. in diameter and $\frac{1}{4}$ in. thick.

As soon as made they were left in dry air, the initial and final set was found to be normal. They were never immersed in water but remained in dry air for several weeks. No cracks occurred and they became so hard and strong that these pats $\frac{1}{4}$ in.

thick, were very difficult to break by the use of the fingers and thumbs. After remaining in dry air for three weeks, they were put out in freezing temperature for three days, and again placed in dry air over the radiator. No cracks or checks have occurred.

After remaining in dry air for a month, a test for absorption was made. A broken pat was weighed dry and found to weigh 94/64 oz. It was then immersed in water for several hours. Upon removal from the water, the surface water was quickly removed with blotting paper, the pat immediately weighed and found to weigh 99/64 oz. Only 5/64 oz. of water was absorbed.

The fact that the pats were never immersed in water and showed no evidence of checking or cracking, and became hard, would indicate that the emulsified oil had held all but the excess water in the mortar and that such mortar was, therefore, both non-evaporative and non-absorbent, which would tend to show that concrete in which mineral oil has been mixed would not be likely to contract and therefore contraction cracks avoided.

Under the theory of Prof. Bauschinger, which has been demonstrated by Prof. Swain in the laboratory of the Institute of Technology, Boston, neat cement when set and hardened in air contracts and that this contraction increases with age up to a certain period, possibly six months or a year. 1 part Portland cement, 3 parts sand hardened in air, shows contraction but less in proportion than neat cement. The results also prove that neat cement when hardened under water shows a slight expansion, while mortar composed of 1 part Portland cement, 3 parts sand, hardened under water, shows expansion, but less in proportion than the neat cement. Reducing these conclusions to figures and taking the average results obtained by various authorities, figuring the expansion and contraction by percentage, the following are the results:

Neat Portland cement hardened in air at the end of 16 weeks shows a 0.15 per cent. contraction.

1 to 3 mortar hardened in air at the end of 16 weeks shows 0.05 per cent. contraction.

Neat Portland cement hardened under water at the end of 16 weeks shows 0.05 per cent. expansion.

1 to 3 mortar hardened under water at the end of 16 weeks shows a 0.015 per cent. expansion.

Mixing oil with concrete from the meager tests so far made, would seem to indicate that the oil held the water in the mortar keeping the cement particles wet and thus furnishing the same conditions as if set under water, hence very materially assisting, if not altogether obviating contraction cracks and hair cracks or crazing. Furthermore, the resulting mortar appears to be far less brittle and therefore such treatment should admirably serve the purposes required of concrete retaining walls, foundations enclosing cellars, tanks, cisterns, etc.

Exhaustive tests have been made by a number of authorities on the action of oils on concrete. The effect of oil on concrete and the effect of oil emulsified in concrete are two separate and distinct subjects.

We are informed by reliable authorities that concrete immersed in animal or vegetable oils will in time disintegrate and that concrete immersed in mineral oils is unaffected. In the first instance there was no chance for the oil to emulsify, in the latter the oil is separated into minute globules. A large field of usefulness is ready for oil mixed and emulsified in concrete. The emulsion takes place after the oil is mixed with the wet concrete and not before, as has been done in a patented article.

A mere casual glance at the uses of Portland cement concrete would indicate that oils mixed with the concrete would prove very desirable for dustless waterproof floors for office buildings, for slaughter house non-absorbent floors, impervious concrete drain tile and sewers. If the experiments to be carried on in the future prove that mineral oils in the course of time are not disadvantageous, the drain tile problem has been solved, for there can be no action of the alkalies or other injurious elements to non-absorbent, dense and impervious concrete.

Such concrete will be practically desirable for silos. Some of the acids formed by the silage in the bottom of the silo would probably not attack a dense, non-absorbent impervious concrete.

Contraction cracks will be eliminated in retaining walls, cisterns, drinking troughs, live stock feeding floors and platforms. Some objection may be raised to the use of oil mixed concrete from the standpoint of its liability to flavor the water or the food. If we stop to consider that the oil is divided into minute globules, thoroughly emulsified, we will see that while there may be some odor there is not likely to be any taste after the drinking trough, feeding floor or cistern has been in use for a few days.

Such oil mixed concrete will be effective for liquid manure cisterns for the reason above described. It will also be particularly adapted to terrazzo floors. The great objection at present being due to contraction cracks. A white oil may be mixed with Portland cement, white sand and water and used for the purpose of setting brick and stone; it being non-evaporative and non-absorbent no efflorescense or stain can occur. In fact such concrete can be used in any work not requiring extraordinary compression strength, and in which the concrete does not come in contact with the heat.

One of the particular advantages will be for stucco work, the exterior plasters.

It would seem that this idea of mixing oil with wet mortar was novel and new, but like many discoveries it only proves to be a re-discovery. In the first century A. D., Marcus Vitruvius Pollio, the famous Roman architect, gives the following detailed specification for stucco. "A mixture of well hydrated lime, marble dust and white sand mixed with water, to which mixture is added either hog's lard, curdled milk or blood."

In A. D. 1280 at Rockingham Castle, England, melted wax was mixed with the mortar.

In A. D. 1324 in the work of King Edward II. at Westminister, pitch was mixed with mortar.

The permanency of the Roman stuccoes may be partially accounted for by the use of oil mixed with mortar. Although Vitruvius used hog's lard, an animal oil, the mortars have withstood the action of the centuries, and in places where freezing temperature occurs in winter and great heat in summer. However the hog's lard must have been very thoroughly emulsified by the action of the hydrated lime. Portland cement was unknown at that period.

In this connection, I would like to suggest the following specification for stucco, the third or finish coat:

One part Portland cement, 20 per cent. (volume of cement) of hydrated lime. 3 parts coarse white sand. First dry mix the sand and cement and with this mix dry hydrated lime, turning each three times with shovels, rake while shoveling. Add water, turning and raking until the desired consistency is obtained. Then add 15 to 20 per cent. of white oil petrole (Chesebrough Mfg. Co., New York), the oil to be by weight in percentage to the weight of the cement. A gallon of oil petrole weighs 7½ pounds. Apply this mortar while the scratch coat is damp and as soon as scratch coat is firm enough to stand the pressure or plastering. If it be desirable to tint the stucco, color the oil with any lime proof coloring matter, in proportion which by experiment with small samples is necessary to give the desired tint.

A white non-volatile mineral oil is suggested for stucco and for mortar to be used in setting white marble or light colored brick, on account of the color possibilities. For concrete where the color is not essential the heavy black bituminous oils to the light non-volatile petroleum oils are successful. They are cheap and their name is legion. Do not use oils containing organic matter and positively avoid, at least for the present and until further experiments have been made, vegetable or animal oils, as they are liable to form an acid which in turn may disintegrate the concrete.

Lime, sand and animal oils have stood the test of centuries; Portland cement and animal oils have not yet had this opportunity. It is within the range of possibility that the test of time may prove contrary to the theory and animal oils emulsified be found not dangerous.

APPENDIX C.

TABLES.

The following tables are included in this volume to present to the reader a brief compendium of present day data. These are mostly the results of laboratory tests, and we can not overlook the opportunity of emphasising the great value of personal field data. We urge upon the reader the necessity of keeping careful and accurate record of field tests and experience, made under his direction or observation. Following is a list of the tables:

TABLE 1.

COMPARATIVE COST OF CONCRETE (All Hand Mixed)

Material, Labor, etc.	*Miami River Bridge, Fernold, Ohio				†Ernst Street Viaduct, Cincinnati, Ohio			
	Quantity used	Price paid	Total cost	Cost per yd. of Concrete in place	Quantity used	Price paid	Total cost	Cost per yd. of Concrete in place
Cement, bbls.	1156	$2 10	$2436 36	$1 58	376	$1 70	$639 00	$1 48
Sand, yds.	....	.....	532 30	35	224	1 20	278 40	64
Stone	....	.....	1165 00	75	255	1 55	433 25	1 00
Lumber	....	.....	981 71	64	....	.....	173 40	40
Tools—Hardware	....	.....	307 18	20	....	.....	25 03	06
Water	....	......	....	.....	....	.....	......	.....
Total Material	....	.....	$5422 55	3 52	....	.....	$1548 65	$3 58
Clearing—Excavating	....	.....	....	.....	....	.....	482 61	1 12
Mixing and Placing Concrete	....	.....	....	.....	....	.....	488 62	1 13
Building Forms, etc	....	.....	....	.....	....	.....	107 37	25
Pumping	....	.....	236 00	15	....	.....	......	.....
Total Labor, per day	....	1 75	4287 86	2 78	....	1 75	1082 20	2 50
Total Cost of Concrete, yd	1542	.....	$9710 41	$6 30	434	.....	$2630 85	$6 08

* One Abutment, 6 river piers, put down with cofferdams. Sand and stone close to site. Cement teamed ten miles.
† Two Abutments at street. Excavation rock and shale.

Material, Labor, etc.	*Quebec Ave. Viaduct, Cincinnati, Ohio				†C. H. & D., B. & O. Viaduct, Cincinnati, Ohio			
	Quantity used	Price paid	Total cost	Cost per yd. of Concrete in place	Quantity used	Price paid	Total cost	Cost per yd. of Concrete in place
Cement, bbls.	500	$1 60	$800 00	$1 40	1908	$1 60	$3052 80	$1 44
Sand, yds.	239	1 25	298 75	53	1105	95	1049 75	50
Stone	560	1 88	1046 90	1 84	1468	1 48	2172 60	1 03
Lumber	....	.....	220 70	38	....	.....	1135 80	54
Tools—Hardware	....	.....	30 44	05	....	.....	537 00	25
Water	....	.....	....	.....	....	.....	50 00	03
Total Material	....	.....	$2396 79	4 20	....	.....	$7797 95	$3 79
Clearing—Excavating	....	.....	....	.....	....	.....	....	.....
Mixing and Placing Concrete	....	.....	....	.....	....	.....	....	.....
Building Forms, etc	....	.....	....	.....	....	.....	....	.....
Pumping	....	.....	....	.....	....	.....	....	.....
Total Labor, per day	....	1 75	1686 00	2 96	....	1 75	7274 67	3 44
Total Cost of Concrete, yd	570	.....	$4062 79	$7 16	2111	.....	$15272 62	$7 23

* All Pedestals 5' x 5' on top and from 8' to 20' high; location very inconvenient for delivery of materials; all team or wheelbarrow work. † One pier 56' high; Concrete handled by steam derrick; two abutments and remainder pedestals; all materials teamed or wheeled.

TABLE 2.

PRICES OF PORTLAND CEMENT TO PRODUCE MORTAR OR CONCRETE OF EQUAL COST TO THAT FROM NATURAL CEMENT AT $1.00 PER BARREL.—(*Taylor & Thompson.*)

Proportions of Natural Cement Mortar	Proportions of Portland Cement Mortar							Proportions of Natural Cement Concrete	Proportions of Portland Cement Concrete				
	1:1	1:1½	1:2	1:2½	1:3	1:3½	1:4		1:2:4	1:2½:5	1:3:6	1:4:8	1:5:10
	$	$	$	$	$	$	$		$	$	$	$	$
1:1	1 00	1 23	1 46	1 69	1 92	2 15	2 38	1:2:4	1 00	1 15	1 32	1 67	2 01
1:1½		1 00	1 18	1 37	1 55	1 74	1 92	1:2½:5		1 00	1 14	1 44	1 72
1:2			1 00	1 15	1 30	1 46	1 61	1:3:6			1 00	1 26	1 51
1;2½				1 00	1 13	1 26	1 39						
1:3					1 00	1 12	1 23						

NOTE.—When the Natural cement is higher or lower than $1.00 per barrel, multiply its cost by the figures in the table to obtain approximate corresponding cost of Portland cement with which it is compared. Values make no allowance for difference in strength or labor of laying mortar.

TABLE 3.

RELATIVE ECONOMY OF DIFFERENT PRICED PORTLAND CEMENTS.—(D. M. Andrews.)

No. of Sample Barrel	Price per Barrel	Relative Cheapness	Fineness		Time of Setting		Tensile Strength						§Relative Strength of 1:3 Mortar	Rel. Economy 1:3 Mortar. Strength X Cheapness	Remarks
					Initial	Final	Neat			1:3 Mortar					
			No. 50 sieve	No. 100 sieve	Hours	Hours	7 days	30 days	60 days	7 days	30 days	60 days	60 days	60 days	
1	$2 77	100.0	93.3	87.6	2	8	324	437	430	66	128	168	79.7	79.7	
2	2 79	99.3	99.3	87.3	8‡	8‡	282	429	468	62	103	124	59.1	58.7	
3	2 82	98.2	98.2	89.7	2	3¼	272	373	431	35	65	87	41.2	40.5	Air pat cracked very slightly
4	2 82	98.2	100.0	99.6	5	9¼	369	460	564	144	184	209	99.4	97.7	
5†	2 89	95.8	99.0	86.2	7¼	7¼	449	543	631	114	175	210	100 0	95.8	
6	2 90	95.5	94.6	77.0	4	6¼	150	227	264	25	57	90	42 8	40.9	Lumpy and gritty on mixing
7	2 98	94.5	100.0	90.0	4	8	440	588	568	127	156	202	96.2	90.9	
8	3 02	91.7	99.5	89.4	2¼	7	418	476	561	89	134	174	82 4	75.6	
9	3 05	90.8	98.5	91.2	3½	7¼	436	518	502	93	126	144	68 2	62.0	Pats cracked slightly
10	3 29	84.2	99.4	92.7	2½	5	365	496	573	78	117	141	67.1	56.5	

†Accepted in preference to No. 4 because air pat slightly defective.
‡Cement not yet set. §Based on the highest. No. 5, as 100.0.

TABLE 4.

PROPORTIONS OF INGREDIENTS ADOPTED IN EUROPE—(MARSH).

Mixture	Quantities necessary to make 1 cubic yard of Concrete			Proportions in cubic ft. of Aggregates per bag of 224 lbs. cement	
	Cement, pounds	Sand, cubic ft., measured loose	Stone, cubic ft., measured loose	Sand, cubic ft.	Stone, cubic ft.
That very generally employed for columns, beams and slabs	505	10.78	21.56	4.31	8.62
Used by some constructors for the same pieces	560	10.78	21.56	4.00	8.00
Employed by M.Considere for hooped Concrete with a resistance of 1,000 lbs. per square inch	560	10.78	21.56	4.00	8.00
Employed by M.Considere for hooped Concrete with a resistance of 1,425 lbs. per square inch and for piles	757	10.78	21.56	2.96	5.92

TABLE 5.

EFFECT OF REGRINDING COARSE PARTICLES AND OF SUBSTITUTING SAND.
(*David D. Butler.*)

Cement, how treated	Fineness residue per cent on sieves of meshes per linear inch			Setting properties		Tensile strength in pounds per square inch									
						Neat cement					1 part cement to 3 parts sand				
	180	76	50	Initial set min.	Final set min.	7 days	28 days	3 months	6 months	12 months	7 days	28 days	3 months	6 months	12 months
As received	33.7	15.5	4.6	13	90	504	580	641	702	717	194	262	354	404	421
Reground	1.3	0.0	0.0	2	20	497	478	518	489	504	326	411	531	591	618
Sand substituted for coarse particles						414	480	606	660	702	164	217	290	354	387

TABLE 6.

QUANTITIES OF MATERIALS FOR ONE CUBIC YARD OF RAMMED CONCRETE, BASED ON A BARREL OF 3.8 CUBIC FEET.—(*Taylor & Thompson.*)

PROPORTIONS BY PARTS			PROPORTIONS BY VOLUMES			Volume of mortar in terms of percentage of volume of stone	PERCENTAGES OF VOIDS IN BROKEN STONE OR GRAVEL														
							50%*			45%†			40%‡			50%§			50%§		
Cement	Sand	Stone	Packed Cement	Loose Sand	Loose Stone		Cement	Sand	Stone	Cement	Sand	Stone	Cement	Sand	Stone	Cement	Sand	Stone	Cement	Sand	Stone
			cu. ft.	cu. ft.			bbl.	cu. yd.	cu. yd.	bbl.	cu. yd.	cu. yd.	bbl.	cu. yd.	cu. yd.	bbl.	cu. yd.	cu. yd.	bbl.	cu. yd.	cu. yd.

NOTE.—Variations in the fineness of the sand and the compacting of the concrete may affect the quantities 10% in either direction.
*Use 50% columns for broken stone screened to uniform size.
†Use 45% columns for average conditions and for broken stone with dust screened out.
‡Use 40% columns for gravel or mixed stone and gravel.
§Use these columns for scientifically graded mixtures.

TABLE 7.

BEHAVIOUR OF PRISMS WHILE SETTING IN AIR.—(*Marsh.*)

Contractions in $\dfrac{1}{100,000}$ of the original length

Number of Days After Molding	1	2	3	4	5	6	7	14	21	28	35	42	49	56	63
Neat Cement } Plain	60	58	57	58	60	64	70	95	110	118	123	128	130	131	132
Reinforced	6	9	12	14	16	17	20	22	23	24	25	25	25	25	25
Mortar } Plain	∽	21	20	21	22	26	29	38	42	44	45	47	47	49	50
Reinforced		6	7	8	9	9	9	9	10	10	10	10	10	10	10

TABLE 8.

BEHAVIOUR OF PRISMS WHILE SETTING UNDER WATER.— (*Marsh.*)

Elongations in $\dfrac{1}{100,000}$ of the original length

Number of Days After Molding	1	2	3	4	5	6	7	14	21	28	35	42	49	56	63
Neat Cement } Plain	7	15	21	27	32	37	41	59	69	73	75	77	78	78	79
Reinforced	2	3	4	5	6	8	9	13	16	18	20	21	22	22	22
Mortar } Plain	3	10	13	15	17	18	19	20	22	24	26	27	27	27	28
Reinforced	2	2	2	3	3	3	4	4	4	4	5	5	5	5	6

TABLE 9.

RESULTS OF TESTS MADE IN 1902 ON EFFECT OF CONTINUOUS MIXING OF
PORTLAND CEMENT MORTARS BY C. G. STREELS, ASSISTANT
CITY ENGINEER, SIOUX CITY.

Sand through sieve of 20 meshes per lin. inch and retained on sieve with 30 meshes per lin. inch; 99 oz. of cement, 220 oz. of sand and 24 oz. of water, or 7.77 per cent of the cement and sand.

No. of Briquettes	Continuously mixed for hrs. min.		Average tensile strength lbs. per sq. inch
4	0	15	294
2	0	30	278
2	0	45	282
2	1	00	243*
2	1	25	275
2	1	55	283
2	2	25	287*
2	2	55	314
2	3	25	326*
2	3	55	372
2	4	25	334†
2	4	55	384
2	5	25	264
2	5	55	249
2	6	25	308*
2	7	25	217
2	8	25	255
4	8	55	246
4	8	55	220*
2	8	55	215*

* 2 oz. water added.   † 3 oz. water added

TABLE 10.

TESTS OF SLAG CEMENT.

Authority	Place where Made	Percentage of Residue	Meshes of Sieve per Lineal Inch	Tensile Strength Pounds per Square Inch								
				Neat					1 to 3			
				Days					Days			
				2	7	28	42	84	7	28	42	84
Pavin de Lafarge	Vitry-le-Francois	15	178		284	455		483	213	355		310
MM. Bergner and Guillerme*				256 to 356	398 to 484	568 to 710	.	710 to 854				
MM. Bergner and Guillerme	Cleveland....	19	127		350	436			240	334	360	396
	Cleveland....	21	127		404	485	502		216	366	396	427
	Newcastle...	16	127		561	700			298	420	456	
	Cleveland....	12	127		483	611	645		289	405	430	450

* Good sample.

TABLE 11.

TESTS OF SLAG CEMENT.
(COMPRESSION.)

Authority	Place where Made	Percentage of Residue	Meshes of Sieve per Lineal Inch	Compressive Strength Pounds per Square Inch								
				Neat					1 to 3			
				Days					Days			
				2	7	28	42	84	7	28	42	84
Pavin de Lafarge	Vitry-le-Francois	15	178		2560	3124		4972	1706	3413		3550
MM. Bergner and Guillerme*				2200 to 2620	3410 to 6130	4770 to 6190		4890 to 7330				
MM. Bergner and Guillerme	Cleveland....	19	127									
	Cleveland....	21	127									
	Newcastle...	16	127									
	Cleveland....	12	127									

* Good sample.

TABLE 12.

RESULTS OBTAINED IN TESTS BY PROF. BACH ON ELASTICITY OF CONCRETE UNDER COMPRESSION.

Proportions of Ingredients in Test Pieces				Value Given by Prof. Bach for Formula $1 - \frac{1}{Ep}$		Values of $c \sim p$ in Pounds per Square Inch									
						114	228	342	456	570	684	798	912	1026	1140
Cement	Danube Sand	Egginge Sand	Broken Stone	Danube Shingle	n	Ep Lbs. per sq. in.	Corresponding Values of $\frac{Ec}{10^6}$ in Lbs. per sq. in. from Formula $Ec = \frac{Ep}{p^{(n-1)}}$								
1	1.5				1.09	4.63x10⁶	2.94	2.77	2.67	2.60	2.58	2.50	2.47	2.44	2.42	2.39
1	3				1.11	6.79x10⁶	4.02	3.73	3.57	3.45	3.36	3.31	3.26	3.20	3.16	3.13
1	4.5				1.15	6.69x10⁶	3.28	2.96	2.79	2.66	2.56	2.50	2.44	2.40	2.56	2.82
1	2.5			5	1.17	5.13x10⁶	2.29	2.08	1.89	1.80	1.72	1.68	1.63	1.61	1.58	1.55
1					1.14	6.16x10⁶	3.17	2.87	2.71	2.62	2.52	2.46	2.42	2.37	2.33	2.30
1		2.5	5		1.16	9.96x10⁶	4.65	4.17	3.91	3.74	3.59	3.49	3.41	3.35	3.28	3.23
1				6	1.14	5.79x10⁶	2.97	2.70	2.54	2.45	2.36	2.32	2.27	2.23	2.19	2.20
1	3		6		1.16	8.23x10⁶	3.87	3.47	3.26	3.J4	2.99	2.91	2.83	2.77	2.73	2.68
1	5			10	1.16	4.73x10⁶	2.22	1.98	1.85	1.78	1.70	1.66	1.62	1.58	1.57	1.56
1	5	10			1.20	8.90x10⁶	3.44	3.00	2.76	2.62	2.49	2.40	2.33	2.27	2.22	2.20

TABLE 13.

ELASTIC PROPERTIES OF CINDER CONCRETE, 12-INCH CUBES AT THREE MONTHS.—WATERTOWN ARSENAL.

American Portland Cement.	Proportions.			Age when Tested.	Modulus of Elasticity between loads per sq. in.			Permanent sets after loads per sq. in. of			Compressive strength lb. per sq. in.
	Cement.	Sand.	Cinder.		100 and 600 lb.	100 and 1000 lb.	1000 and 2000 lb.	600 lb.	1000 lb.	2000 lb.	
A	1	1	3	90	2 500 000	2 500 000	1 429 000	0.	.0001	.0006	2 780
	1	2	5	90	1 087 000	957 000			.0008	.0028	1 402
	1	2	5	90	1 471 000	1 286 000			.0002	.0010	1 715
B	1	1	3	90	4 167 000	3 214 000	1 190 000	0.	.0001	.0014	2 368
	1	1	3	90	2 083 000	1 875 000	1 351 000	.0001	.0002	.0017	2 580
	1	2	5	90	1 190 000	849 000			.0009	.0066	1 200
	1	2	5	90	1 087 000	865 000			.0024	.0089	1 263

*Tests of Metals, U. S. A., 1898, pp. 561 and 573.

TABLE 14.

ELASTIC PROPERTIES OF BROKEN STONE CONCRETE, 12-INCH CUBES. PORT-
LAND CEMENT,* BANK SAND AND BROKEN CONGLOMERATE STONE.
BY GEORGE A. KIMBALL AT WATERTOWN ARSENAL.

COMPOSITION				MODULUS OF ELASTICITY BETWEEN LOADS PER SQUARE INCH OF			Compressive strength per sq. in.
Cement	Sand	Broken Stone	Age	100 and 600 lb.	100 and 1 000 lb.	1 000 and 2 000 lb.	lb.
1	2	4	7 days	2 593 000	2 054 000	1 351 000	1 730
1	2	4	1 mo.	2 662 000	2 445 000	1 462 000	2 567
1	2	4	3 mos.	3 671 000	3 170 000	2 158 000	2 975
1	2	4	6 mos.	3 646 000	3 567 000	2 582 000	3 989
1							
1	3	6	7 days	1 869 000	1 530 000		1 511
1	3	6	1 mo.	2 438 000	2 135 000	1 219 000	2 260
1	3	6	3 mos.	2 976 000	2 656 000	1 805 000	2 741
1	3	6	6 mos.	3 608 000	3 503 000	1 868 000	3 068
1							
1	6	12	1 mo.	1 376 000			1 146
1	6	12	3 mos.	1 642 000	1 364 000		1 359
1	6	12	6 mos.	1 820 000	1 522 000		1 592

TABLE 15.

COMPRESSIVE STRENGTH OF 12-INCH CUBES OF CINDER CONCRETE.—WATER-
TOWN ARSENAL.

Cement.	Proportions Cement. Sand. Cinder.			Age, 1 month.		Age, 3 months.	
	Cement.	Sand.	Cinder.	Mean weight lb. per cu. ft.	Compressive strength lb. per sq. in.	Mean weight lb. per cu. ft.	Compressive strength lb. per sq. in.
German Portland.............	1	1	3	112.1	1 466	110.4	2 001
	1	2	3	115.2	1 098	112.8	1 634
	1	2	4	111.2	904	107.9	1 325
	1	2	5	108.8	769	105.3	1 084
	1	3	6	107.6	529	103.5	788
American Portland............	1	1	3	117.2	1 965	115.2	2 624
	1	2	5	111.3	818	110.0	1 412

TABLE 16.

ADHESIVE TESTS OF DEFORMED BARS MADE BY PROF. SIBLEY AT CASE SCHOOL OF APPLIED SCIENCE.

Type of Bar	No. of Tests	Size of Bar In. diam.	Area Cross Section Sq. In.	Perimeter Inches	Length Embedded	Area Embedded	Average Load				Av. Unit Bond at Start Slip	Av. Unit Bond Max. Load	Remarks
							Start Slip	.006 Slip	.0125 Slip	Max. Load			
Twisted Lug.........	3	½	.23	1.8	5.95	10.71	3675	6490	6710	6710	343	626	Clean.
" "	3	¾	.54	2.83	7.83	22.17	8840	11240	11325	11325	399	511	Very Rusty.
Ransome............	3	½	.24	1.94	5.93	11.52	3190	4270	4615	5310	272	461	Clean; 10 spirals in 8″
"	3	¾	.56	2.90	7.82	22.65	7970	8850	9990	9070	352	401	Clean; 5 spirals in 7″
Thacher.........	3	½	.23	1.83	5.98	10.96	4230	.....	6180	6290	386	575	Clean.
"	3	¾	.42	2.57	7.93	20.38	5380	8480	8960	8960	264	441	Clean.
Diamond............	3	½	.26	1.91	5.96	11.39	4825	7160	7300	7300	423	641	Clean.
"	3	¾	.58	2.85	8.57	24.41	9670	11980	12700	12860	396	525	Clean.
New Style Johnson.	3	½	.24	1.94	5.97	11.56	5510	6870	6870	6870	477	598	Very Rusty.
Old Style Johnson..	4	¾	.55	2.92	7.90	23.06	8950	10360	10360	10360	388	448	Very Rusty.

TABLE 17.

WARREN'S TESTS ON ADHESION OF CONCRETE TO STEEL.

Description	Composition				Age in days	Adhesion in pounds per square inch of surface
	Cement	Sand	¾ Shivers	Water per cent.		
Bars with natural skin on Hardened in air	1 1 1 1	3 3 2 2	2 2	12 12 10 10	45 45 45 45	216.5 221.0 184.5 170.0 } mean 198
Bars cleaned with emery paper Hardened in air	1 1 1 1	3 3 2 2	2 2	12.5 12.5 10 10	45 45 44 44	118.0 72.0 154.0 155.0 } mean 125
Bars cleaned with emery paper Hardened in water	1 1 1 1	3 3 2 2	2 2	12 12 10 10	45 45 45 45	154.0 191.0 204.0 191.0 } mean 185

TABLE 18.

TYPICAL ANALYSIS OF CEMENTS—(*Taylor & Thompson.*)

	Portland cement		Natural Cement						Puzzolan Cement	Hydraulic Lime (Le Tiel)	Common Lime	
			American		Eng.	French						
	Lehigh Valley (mixed rock)	Western (marl and clay)	Eastern Rosendale	Western Louisville	Roman	Vassy	Grapplers			Lime	Magnesian Lime
Silica Si O₂	21.31	21.93	18.38	20.42	25.48	22.60	26.5	28.96	21.70	1.08	1.12
Alumina Al₂ O₃	6.89	5.98	15.20 {	4.76	10.30	8.90	2.5	11.40	3.19	1.27 {	0.69
Iron Oxide Fe₂ O₃	2.53	2.35		3.40	7.44	5.30	1.5	0.54	0.66		
Calcium Oxide Ca O	62.89	62.92	35.84	46.64	44.54	52.69	63.0	50.29	60.70	97.02	58.52
Magnesian Ox. Mg O	2.64	1.10	14.02	12.00	2.92	1.15	1.0	2.96	0.85	0.68	39.58
Sulphuric Acid S O₃	1.34	1.54	0.93	2.57	2.61	3.25	0.5	1.37	0.60		
Loss on Ignition	1.39	2.91	3.73	6.75	3.68	6.11	5.0	3.39	12.20		
Other constituents	0.75		11.46	8.74	1.46			0.30	0.10		

TABLE 19.

Percent of Water for Cement Mortars of Normal Consistency.
(Suggested by American Society for Testing Materials.)

Percentage of Water for Neat Cement	Percentage of Water for Sand Mortars					Percentage of Water for Neat Cement	Percentage of Water for Sand Mortars				
	Proportions Cement to Sand by Weight						Proportions Cement to Sand by Weight				
	1:1	1:2	1:3	1:4	1:5		1:1	1:2	1:3	1:4	1:5
18	12.0	10.0	9.0	8.4	8.0	33	17.0	13.3	11.5	10.4	9.6
19	12.8	10.2	9.2	8.5	8.1	34	17.3	13.6	11.7	10.5	9.7
20	12.7	10.4	9.3	8.7	8.2	35	17.7	13.8	11.8	10.7	9.9
21	13.0	10.7	9.5	8.8	8.3	36	18.0	14 0	12.0	10.8	10.0
22	13.3	10.9	9.7	8.9	8.4	37	18.3	14.2	12.2	10.9	10.1
23	13.7	11.1	9.8	9.1	8.5	38	18.7	14.4	12.3	11.1	10.2
24	14.0	11.3	10.0	9.2	8.6	39	19.0	14.7	12.5	11.2	10.3
25	14.3	11.6	10.2	9.3	8.8	40	19.3	14.9	12.7	11.3	10.4
26	14.7	11.8	10 3	9.5	8.9	41	19.7	15.1	12.8	11.5	10 5
27	15.0	12.0	10.5	9.6	9.0	42	20.0	15.3	13.0	11.6	10.6
28	15.3	12.2	10.7	9.7	9.1	43	20.3	15.6	13.2	11.7	10.7
29	15.7	12.5	10.8	9.9	9.2	44	20.7	15.8	13.3	11.9	10.8
30	16.0	12.7	11.0	10.0	9.3	45	21.0	16.0	13.5	12.0	11.0
31	16.3	12.9	11.2	10.1	9.4	46	21.3	16.1	13.7	12.1	11.1
32	16.7	13.1	11.3	10.3	9.5						

TABLE 20.

Effect of Fineness of Sand upon 1 to 2 Cement Mortar.

Size of Sand		Tensile Strength Pounds per Square Inch at the Following Times After Mixing				
Passed Meshes per Lineal inch	Retained Meshes per Lineal inch	After 7 Days	1 Month	3 Months	6 Months	1 Year
4	8	243	442	539	470	668
8	16	269	345	473	512	572
16	20	186	250	313	397	392
20	30	211	281	322	402	440
30	50	149	205	238	275	318
50	75	122	214	250	275	306
75	100	98	153	211	208	253
100	...	98	155	161	229	271

TABLE 21.

Average Specific Gravity of Various Aggregates.

Material.	Specific Gravity.	Weight of a solid cu. ft. of rock. lb.	Authority
Sand	2.65	165	Allen Hazen
Gravel	2.66	165	A. E. Schutté
Conglomerate	2.6	162	Robert Spurr Weston
Granite..................	2.7	168	Edwin C. Eckel
Limestone	2.6	162	Edwin C. Eckel
Trap	2.9	180	Edwin C. Eckel
Slate	2.7	168	Tod's Tables‡
Sandstone	2.4	150	Edwin C. Eckel
Cinders (bituminous)	1.5	95	The authors

‡Encyclopedia Britannica.

TABLE 22.

Percentage of Voids Corresponding to Different Weights per Cubic Foot of Sand, Gravel, and Broken Stone Containing Various Percentages of Moisture.

Weight of one cu. ft. of sand or gravel.†	Percentages of absolute voids in material containing moistures by weight.‡					Moisture by volume corresponding to 1% by weight.‡	Weight of one cu. ft. of sand or gravel.†	Percentages of absolute voids in material containing moistures by weight.‡					Moisture by volume corresponding to 1% by weight.‡
	0%	2%	4%	6%	8%			0%	2%	4%	6%	8%	
	%	%	%	%	%	%		%	%	%	%	%	%
70	57.6	58.4	59.3	60.1	61.0	1.1	98	40.6	41.8	43.0	44.2	45.3	1.6
75	54.5	55.4	56.4	57.3	58.2	1.2	99	40.0	41.2	42.4	43.6	44.8	1.6
80	51.5	52.5	53.4	54.4	55.4	1.3	100	39.4	40.6	41.8	43.0	44.2	1.6
81	50.9	51.9	52.9	53.9	54.8	1.3	101	38.8	40.0	41.2	42.5	43.7	1.6
							102	38.2	39.4	40.7	41.9	43.1	1.6
82	50.3	51.3	52.3	53.3	54.3	1.3	103	37.6	38.8	40.1	41.3	42.5	1.6
83	49.7	50.7	51.7	52.7	53.7	1.3	104	37.0	38.2	39.5	40.8	42.0	1.7
84	49.1	50.1	51.1	52.2	53.2	1.4	105	36.4	37.6	38.9	40.2	41.4	1.7
85	48.5	49.5	50.6	51.6	52.6	1.4	106	35.8	37.0	38.3	39.6	40.9	1.7
86	47.9	48.9	50.0	51.0	52.0	1.4	107	35.2	36.4	37.7	39.0	40.3	1.7
87	47.3	48.3	49.4	50.4	51.5	1.4	108	34.6	35.9	37.2	38.5	39.7	1.7
88	46.7	47.7	48.8	49.9	50.9	1.4	109	33.9	35.3	36.6	37.9	39.2	1.7
89	46.1	47.1	48.2	49.3	50.4	1.4	110	33.3	34.7	36.0	37.3	38.7	1.8
90	45.5	46.5	47.6	48.7	49.8	1.4							
91	44.8	45.9	47.0	48.2	49.2	1.5	115	30.3	31.7	33.1	34.5	35.9	1.8
92	44.2	45.4	46.5	47.6	48.7	1.5	120	27.3	28.7	30.2	31.6	33.1	1.9
93	43.6	44.8	45.9	47.0	48.1	1.5	125	24.2	25.8	27.3	28.8	30.3	2.0
94	43.0	44.2	45.3	46.5	47.6	1.5	130	21.2	22.8	24.4	25.9	27.5	2.1
95	42.4	43.6	44.7	45.9	47.0	1.5	135	18.2	19.8	21.4	23.1	24.7	2.2
96	41.8	43.0	44.1	45.3	46.4	1.5	140	15.2	16.8	18.5	20.2	21.9	2.2
97	41.2	42.4	43.6	44.7	45.9	1.6							

*The weight per cubic foot of a solid is the specific gravity of the rock multiplied by the weight of a cubic foot of water.

†Also applicable to broken stones such as granite, conglomerate, and limestone, whose specific gravity averages from 2.6 to 2.7. Table is based on specific gravity of 2.65.

‡The per cent. of absolute voids given in the columns include the space occupied by both the air and the moisture. To determine the per cent. of air space, multiply the figure in the last column, opposite the weight of sand under consideration, by the per cent. of moisture by weight, and deduct result from the per cent. already found.

INDEX

INDEX

\

INDEX

INDEX

INDEX

INDEX

INDEX

A Concrete Factory Building, Finished. The Frontispiece
Shows This Structure During Erection.

Lightning Source UK Ltd.
Milton Keynes UK
28 March 2011

170019UK00001B/24/P